Process Development Division

Presentations at the 2009 AIChE Annual Meeting

Nashville, Tennessee, USA
8-13 November 2009

ISBN: 978-1-61567-935-5

Printed from e-media with permission by:

Curran Associates, Inc.
57 Morehouse Lane
Red Hook, NY 12571

CURRAN ASSOCIATES INC. proceedings.com

Some format issues inherent in the e-media version may also appear in this print version.

Printed by Curran Associates, Inc. (2010)

AIChE
3 Park Avenue
New York, NY 10016-5991

Phone: (203) 702-7660
Fax: (203) 775-5177

www.aiche.org

Additional copies of this publication are available from:

Curran Associates, Inc.
57 Morehouse Lane
Red Hook, NY 12571 USA
Phone: 845-758-0400
Fax: 845-758-2634
Email: curran@proceedings.com
Web: www.proceedings.com

TABLE OF CONTENTS

A Duck and An Engineer Walk Into a Bar (Stories from Engineering)

Richard F. Garnett, Honda Manufacturing of Alabama, Lincoln, AL

A wise man said, "Learn from the mistakes of others. You cannot live long enough to make them all yourself." Presented in a humorous style this talk shares some of the mistakes made and lessons learned in a twenty-five year career in engineering and project management.

Discussion Points:

- If Only Operators Made Mistakes (In Pursuit of Engineering Perfection)

- What You Get For Your Degree (Am I a Genus Now?)

- Where the Rubber Meets the Road (Designing for People)

- Knowing What it is All About (Money the Optimum Design Factor)

- Trends in Engineering (The Times They Are a Changing)

The Presentation deals with the things that young engineers need to learn to transition from college into industry. Using stories and anecdotes, it discusses the "soft side skills" required to be successful.

See more of this Session: Best Practices in Process Development
See more of this Group/Topical: Process Development Division

Scale-Up of First-of-a-Kind Processes

From Pilot Plant to Commercial Production

AIChE Annual General Meeting. Nashville, TN.
Monday, November 9, 2009.

Anthony Cato

Osman Aboul-Nasr*
Fluor

Greenville, SC. USA

FLUOR®

Disclaimer

Fluor makes no representation or warranty, expressed or implied, as to the accuracy or completeness of the information contained herein and nothing contained herein is, or shall be relied upon as a promise or representation to be used in connection with any proxy, proxy statement, proxy soliciting materials, prospectus, Securities Registration Statement or similar document without the express written consent of Fluor, except as may be required by law.

FLUOR®

Fluor Corporation

- One of world's largest, publicly owned engineering, procurement, construction, and maintenance companies
- *Engineering News-Record (ENR)* magazine consistently ranks Fluor Corporation among the top three on its "The Top Design-Build Firms" and "The Top 100 Contractors by New Contracts" lists
- An international workforce of more than 30,000 employees worldwide
- Founded on a legacy of delivering world-class solutions for almost 100 years
- One of the world's safest contractors

FLUOR®

Fluor's Locations

MKTAV07_STD007.AI 10/07

FLUOR®

Comprehensive Services

Program/Project Management

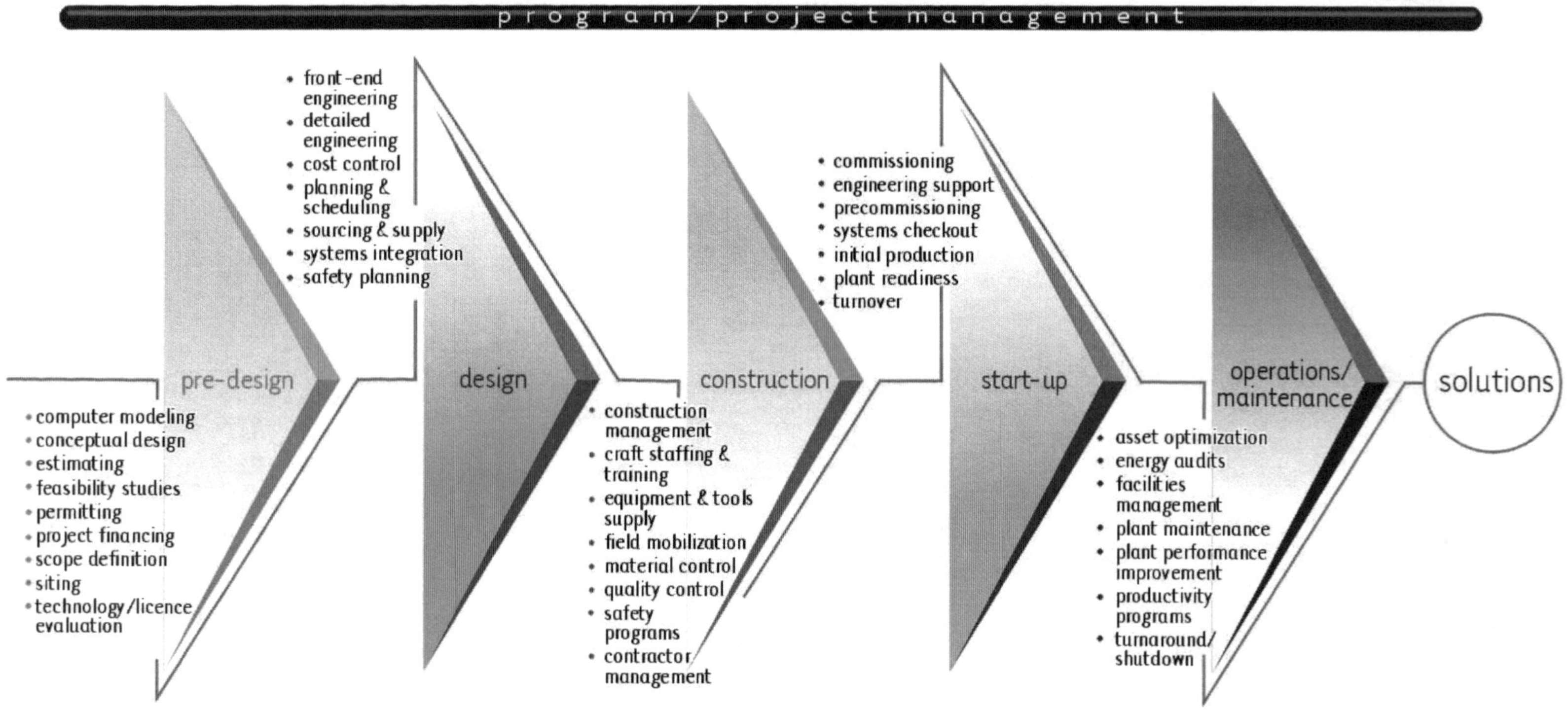

FLUOR®

Outline

- Risks of First-of-a-Kind Technology
- Piloting and Scaling-up from Lab
- Scaling from Pilot to (semi-)commercial
- The Engineering Process
- Lessons Learned

FLUOR®

Business Risks of New Technology

- Cost growth
- Commercial risk
- Downplayed project risks at authorization
- Engineering Risks
- Unpredictable startups
- Technical advantages may impact operability
- Long-term operability problems
- *Failure to Recognize Risk*

FLUOR®

Distinguishing Characteristics

- A Champion
- Need senior management commitment
- Technically driven
- A higher level of technical resources
- Team integration and continuity
- Much more business involvement
- Willingness to spend money to keep pilot running

FLUOR®

Engineering Risks of New Technology

- **Failure to recognize risks**
- **Longer engineering time**
- **Inordinate haste in project execution**
  - Shortcut R&D
  - Short-cut FEL
  - Accelerate execution
- **Not the average engineering project**

FLUOR®

Scaling Up (is hard to do)

- "Make your mistakes on a small scale; earn your rewards on a large scale."
- Choose a realistic scale-up factor.
- Keep commercial operations in mind.

But

- Plugging might be easier to handle.
- Less heat losses
- etc.

FLUOR®

Simulation

- Define the physical properties
- Minor and trace components
- Identify limitations in simulators.
- Do not prove single unit operations separately.
- Include abnormal conditions too, significant impact on safety
- The need to freeze the simulation is a fact of life.

FLUOR®

Recycle Streams

- **Not lab grade products**
- **Concentrations of impurities will rise over time**
  - Unsteady state simulation?
- **Chemical Reaction Matrix**
- **Azeotropes**
- **Materials of construction**

FLUOR®

Into the Pilot Plant

- Small inefficiencies in the pilot plant are amplified on the commercial scale
- Pay special attention to unusual unit operations.
- Identify where scale up will be required.
- Can you really live with off-the shelf items?
- Include large buffer surge vessels.
- Include an operations-savvy member in the team.

FLUOR®

Not of Marginal Importance

- Overinstrument, oversample, overanalyze the pilot plant and data. Do not overconservatize.
- Push the envelop in design and operations
- Use large margins in flows, sizes, numbers, etc.
- Conservative pilot plant design generally leads to expensive commercial production

FLUOR®

Enter the Engineering Company

- Prepare and update PFD and P&ID
- "Make mistakes on a small scale…"
- It is a TEAM EFFORT; take advantage of strength
- Proof of concept is alive and well… throughout.
- Value Engineering is …invaluable.
- The need to freeze pilot runs…
  - Operate to generate data not to ameliorate process.

FLUOR®

Last but not Least

- Pay attention to auxiliary systems (40% of ISBL)
- Layout may be a full-scale challenge.
- Pay attention to distance / close connectivity.
- Include your commercial scale intelligence.
- Operating team involvement is even more essential than normal projects.
- Commercial suppliers begin to appear.
- Identify special / long lead equipment.

FLUOR®

SHE is Important

- Safety becomes more important; not only due to scale.
- Health, toxicology and
- Environmental impact and waste disposal are exponentially more important.

FLUOR®

Lessons Learned

- It is a team effort
- Look at the Bright Side of Life: What Went Wright
- The first project identify more lessons to learn

FLUOR®

Facilitating Process Development and Technology Transfer with Computer Aids

Demetri Petrides, INTELLIGEN, INC., Scotch Plains, NJ

Charles Siletti, INTELLIGEN, INC., Mount Laurel, NJ

The successful scale up and commercialization of pharmaceutical APIs and products is a challenging task that requires collaboration of professionals from many disciplines. Process simulators and other computer aids can facilitate this task by assisting scientists and engineers to answer the following and other related questions: What is the impact of product titer increase on the capacity load of the downstream section, the overall throughput of a plant, and the cost of goods? What changes are required in an existing multi-product facility to accommodate the process of a new product? What is the range of variability that a process can accommodate if it operates under a tight cycle time? What is the impact of single-use equipment on the demand for utilities and the cost of goods? Our experience in addressing the above and other related questions will be presented using industrial examples spanning production of APIs and pharmaceutical products.

See more of this Session: Best Practices in Process Development
See more of this Group/Topical: Process Development Division

Work-Sharing during Design and Construction of Pilot Plants

AIChE Annual General Meeting. Nashville, TN.
Monday, November 9, 2009.

Krystal A. Donald,
Paul E. Samuels and
Osman Aboul-Nasr*
Fluor
Greenville, SC. USA

FLUOR®

Disclaimer

Fluor makes no representation or warranty, expressed or implied, as to the accuracy or completeness of the information contained herein and nothing contained herein is, or shall be relied upon as a promise or representation to be used in connection with any proxy, proxy statement, proxy soliciting materials, prospectus, Securities Registration Statement or similar document without the express written consent of Fluor, except as may be required by law.

FLUOR®

Fluor Corporation

- One of world's largest, publicly owned engineering, procurement, construction, and maintenance companies
- *Engineering News-Record (ENR)* magazine consistently ranks Fluor Corporation among the top three on its "The Top Design-Build Firms" and "The Top 100 Contractors by New Contracts" lists
- An international workforce of more than 30,000 employees worldwide
- Founded on a legacy of delivering world-class solutions for almost 100 years
- One of the world's safest contractors

FLUOR®

Global Presence

Fluor serves customers through a network of offices in more than 25 countries across six continents.

- Argentina
- Australia
- Azerbaijan
- Brazil
- Canada
- Chile
- China
- England
- India
- Indonesia
- Japan
- Mexico
- Netherlands
- Peru
- Philippines
- Poland
- Russia
- Saudi Arabia
- Singapore
- South Africa
- South Korea
- Spain
- United Arab Emirates
- United States of America
- Venezuela

FLUOR®

Services

Fluor provides appropriate, cost-effective solutions in a timely manner for customers worldwide.

Conceptual Design

Engineering

Process Technology

Procurement

Construction

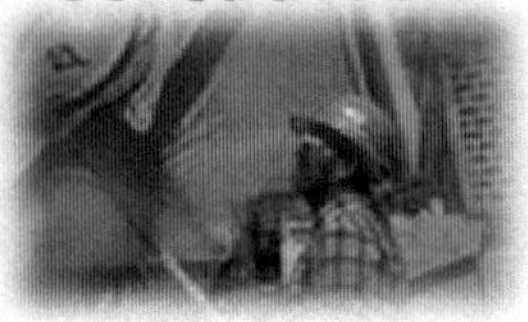

Operations and Maintenance

Project and Program Management

Decommissioning

Facility and Property Reuse

Staffing

FLUOR®

Outline

- The Players
- Engineering Steps
- CAD Details
- Lessons Learned

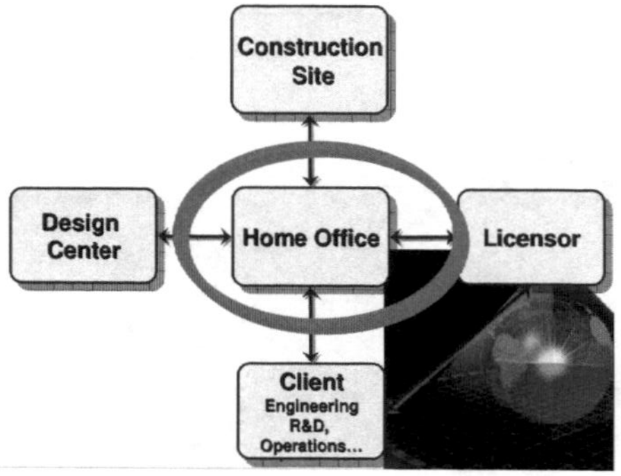

FLUOR®

Sharing the Work

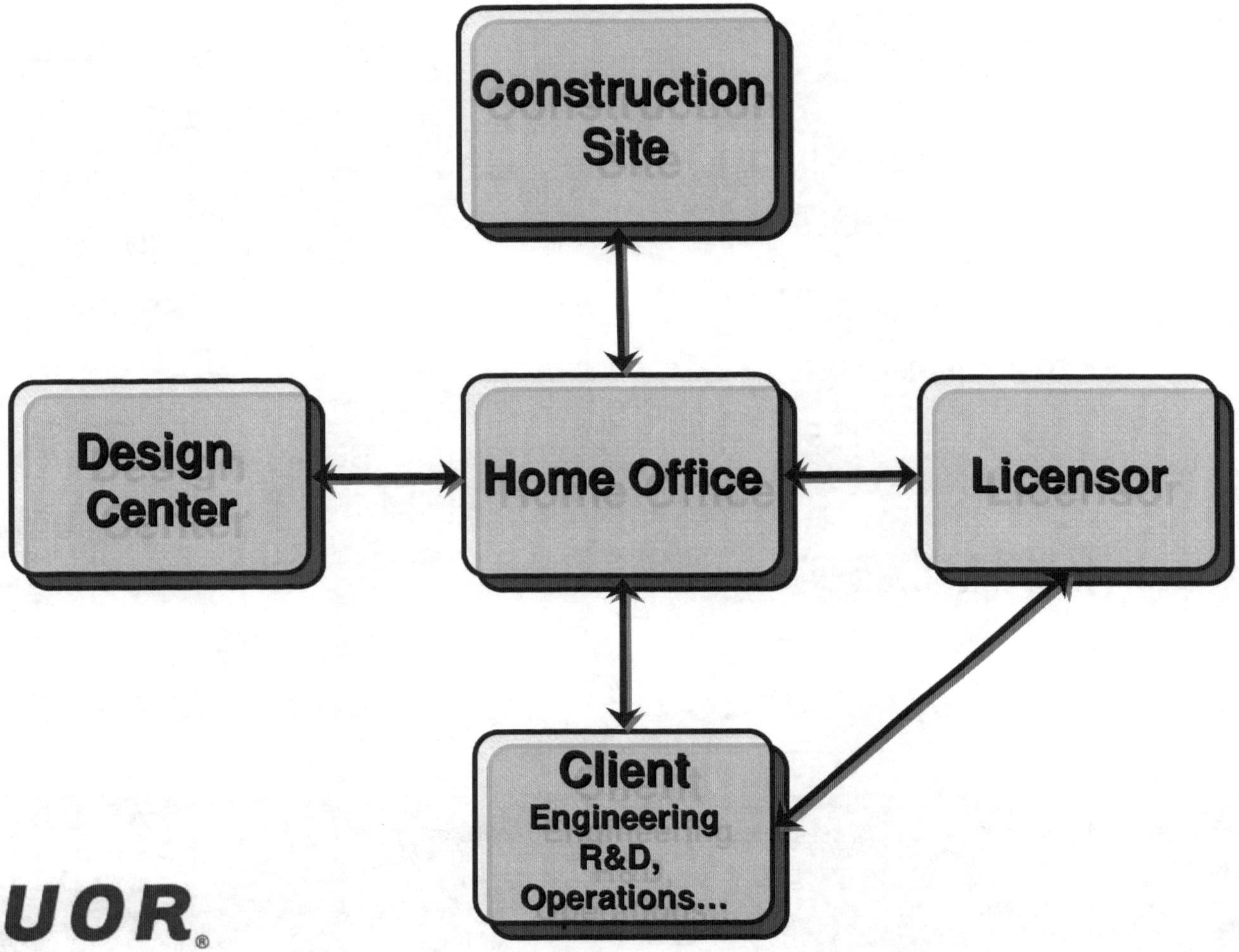

FLUOR®

The Players

- **Engineering Company**
  - Process (Chemical)
  - Project
  - CAD
  - Other Disciplines
- **Owner**
  - Process
  - Project
  - R&D
  - Operations
  - Engineering
  - etc.
- **Design Center**
  - CAD
  - Other disciplines.
- **Site**
  - Project
  - Construction, etc.
- **Licensor**
  - Process
  - Engineering

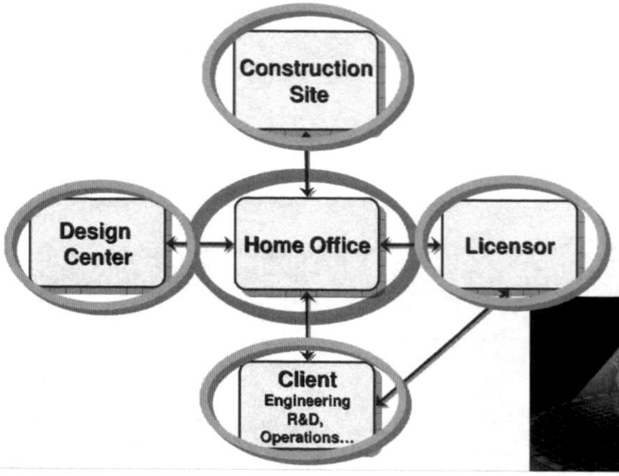

FLUOR®

Project Setup, Technology Transfer

- **Client and Engineering Company**
- **Plan**
  - Personnel
  - Communication Protocols and Procedures
  - Roles & Responsibilities
  - LANGUAGE
  - Technology Exchange
- **Potential Licensor Involvement**

FLUOR®

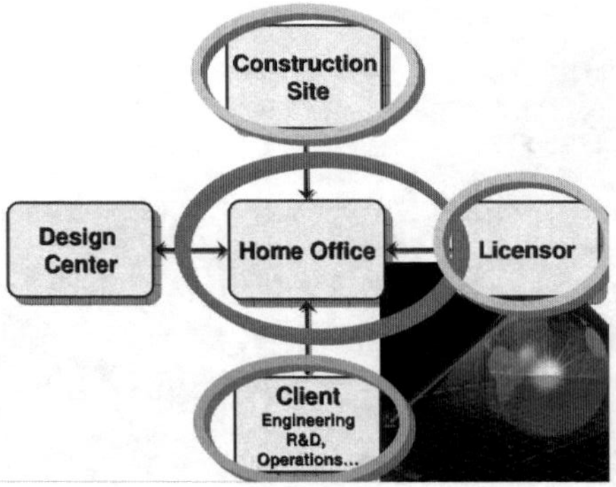

Project Setup, Engineering Setup

- **Engineering Company**
- **Plan**
  - Personnel
  - Communication Protocols and Procedures
  - Roles & Responsibilities
  - LANGUAGE
  - Technology Exchange
- **Global Execution Centers**

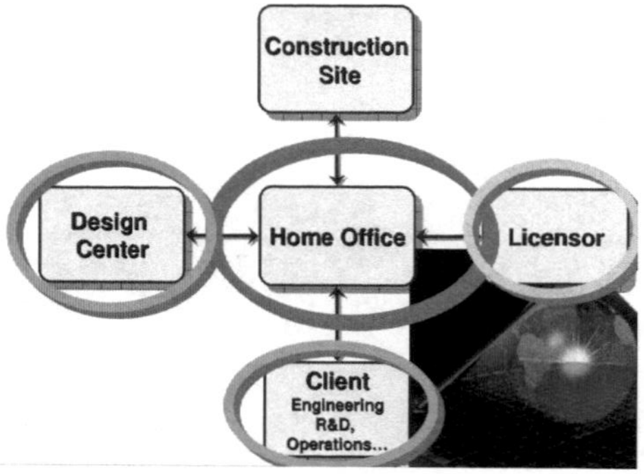

FLUOR®

Engineering Setup

- **Alignment within / between Home Office (s)**
- **Vertical and Horizontal Split**
  - Uniform execution
  - Play to your strengths
- **Formats**
- **Sharing of Databases**
- **Travel**

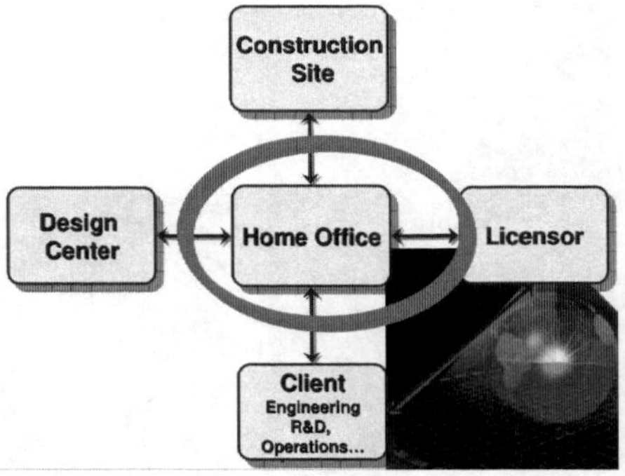

FLUOR®

Engineering Execution

- **Allow for Client / Licensor Interaction**
- **Scheduled Communication**
  - Face-to-Face Meetings
  - Video Conferences
- **Home Office (s)**
  - Single Management/Lead at Every Level
- **Document Control**
  - Squad checks

FLUOR®

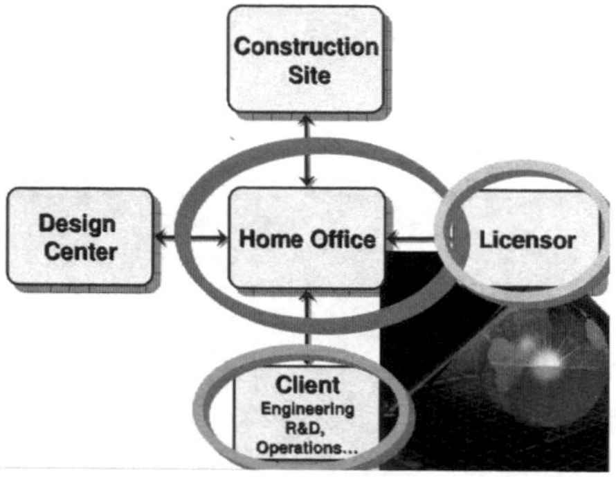

CAD Setup

- **Automation Plan**
  - Project Standards
  - Roles & Responsibilities
  - Software
  - Procedures
  - Backup Plans
- **Workload Distribution**
- **Setup Meeting**

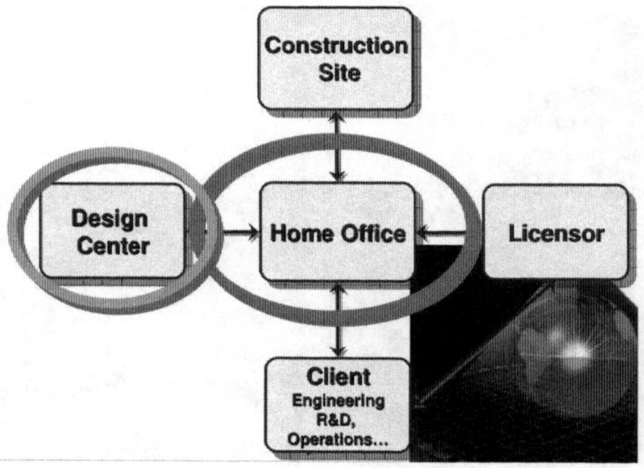

FLUOR®

CAD Work-Share Setup

- Identify Scope, Schedule and Expectation
- Identify / Develop Communication Procedures, Protocols and Portal
- Take Advantage of, and PLAN FOR, different time zones

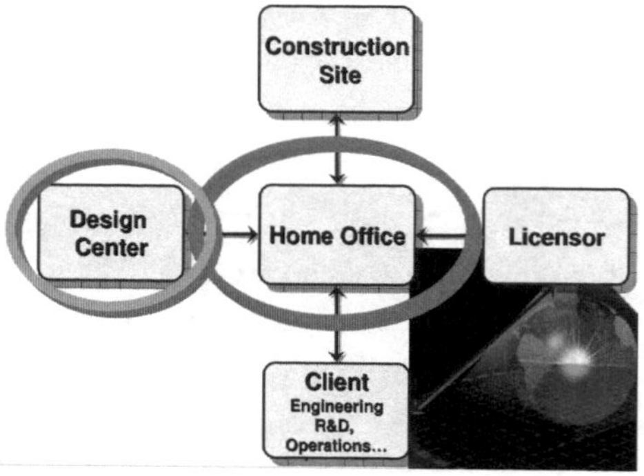

FLUOR®

CAD Start-Up

- **Checklist**
  - Standards
  - Roles & Responsibilities
  - Software
- **Startup Meeting**

FLUOR®

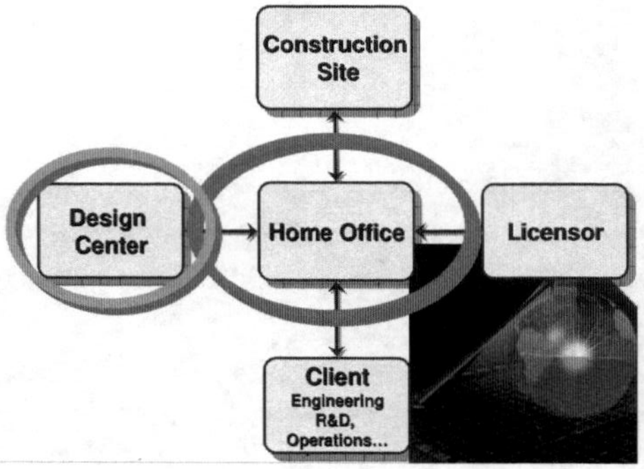

CAD Execution

- **Communication**
  - Frequency of Meeting
  - Procedures, Protocol and Portal
  - Workload Distribution
  - Changes in Standard (Change Management)
- **Methodology**
  - Progress Tracking
  - Workload Distribution
  - Problem Resolution

FLUOR®

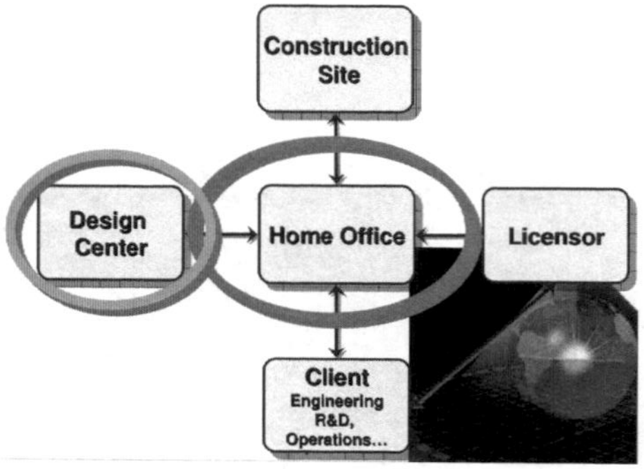

Lessons Learned

- **Talk and Listen**
- **Look at the Bright Side: What Went Wright**
- **Turn Lemon into Lemonade**
  - Opportunities Identified
  - Opportunities Missed
  - Conflicts Faced and Resolution Mechanism
- **Play to Your Strengths**

FLUOR®

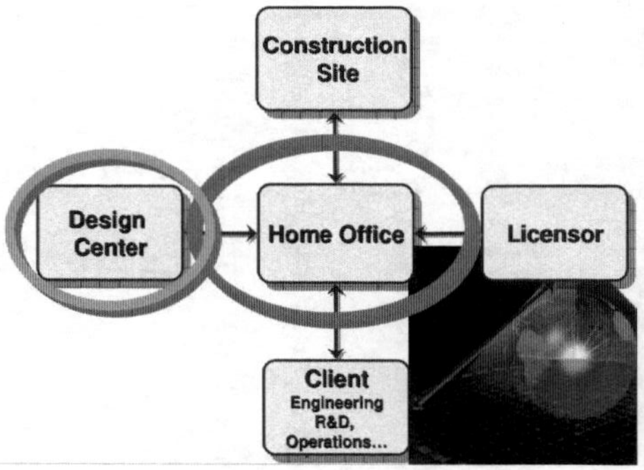

Effective Technology Transfer of Pharmaceutical Capsule Manufacturing

John F. Peragine, Bristol-Myers Squibb Company, New Brunswick, NJ

A pharmaceutical capsule is currently produced at two different manufacturing sites, one in the US and the other in Europe. Since manufacturing at both sites is being discontinued, production needs to be transferred to another site in the US. The manufacturing processes at both current sites, although similar, have some distinct differences in the way the pharmaceutical powder is granulated prior to filling the capsule (encapsulation). Since the new site will be the only site for worldwide production, the new manufacturing process needs to be harmonized (demonstrating that the new process is equivalent to both current processes). Process equivalence is also important to demonstrate for the required FDA and multinational filings before the new site can be approved. This paper discusses the key aspects of the powder granulation and encapsulation unit operations and how sound technology transfer principles are used to manufacture an equivalent capsule product.

See more of this Session: Best Practices in Process Development
See more of this Group/Topical: Process Development Division

Proposal and Validation of A Colombian Biodiesel Plant Simulation, Operative and Economic Optimization BASED ON Numeric Modelling

Alejandra Manrique, Chemical Engineering Department, Universidad de los Andes, Bogotá, Colombia

Maria Alejandra Acero, Chemical Engineering, Universidad de los Andes, Bogotá, Colombia

Federico Velosa, Chemical Engineering, Universidad de los Andes, Bogotá, Colombia

Jorge Mario Gomez, Chemical Engineering, Universidad de los Andes, Bogotá, Colombia

This work validates the numeric simulation carried out of typical Colombian Biodiesel production plants. Based on this validation achieved through its comparison with real operative data of national plants, an optimization of operative conditions is developed using two objective functions: an economic one that seeks to maximize economic yields, and another one in which increasing biodiesel productivity is intended. Finally, an economic analysis is done for the operation and construction of the optimized plant

The simulation involves rigorous mathematical models, considering transesterification reaction kinetics approximated to the kind of oil used in the country [1], and it is based on the plant configuration proposed by Zhang et al [2], taking into account as raw materials waste cooking oil and vegetal oil, and employing acid and alkali catalysts for biodiesel production. From this study, Velosa [3] developed and optimized a simulation at the conditions and characteristics proposed from palm oil. The results were employed as base for the adjustment of the simulation comparing it with an industrial level Colombian process. For its development, data supplied by CORPODIB® (Corporación para el Desarrollo Industrial de la Biotecnología y Producción Limpia) [4] -institution that counts with a biofuels production pilot plant in Bogotá- was used.

After comparing the theoretical process with the industrial, the main differences obtained revolved around the reactors operative conditions and glycerol purification section, subproduct of the reaction. In the adjusted simulation the reactors operate at atmospheric conditions (industrial conditions) instead of the pressures reported by the literature, and the glycerol purification configuration, which consisted of a distillation tower, was replaced by two evaporators. Initially an assessment of the raw materials required was made based on previous research that assures the project viability within the industrial context of the country and the technical specifications of the product. Hence, it was established that methanol and NaOH 1%wt (relative to the oil quantity) must be employed for the transesterification reaction, and HCl must be used in order to remove catalyst from the process.

Once the adjustement of real operative conditions was made, the next step was to carry out the optimization of the working conditions of the plant, trying to maximize two separate objective functions: i) biodiesel production and ii) economic profit taking into account both the capital and manufacture cost. In both cases the configuration Reactor-Separator proposed by CORPODIB was assessed for the transesterification section. This arrangement consists of implementing a decanter after each reactor in order to achieve the separation of biodiesel from glycerol, instead of the original proposal that establishes the Reactor-Reactor sequence without intermediate separation. When the Reactor-Separator simulation was done it was found that it is necessary to add more raw material into the process, and even so a considerable diminishment of FAME (Fatty Acid Methyl Esters) and glycerol production appeared. Due to the previous reasons and as suggestion obtained from the optimization, the original Reactor-Reactor configuration was selected since it grants greater production flows.

The optimization comprehends the three fundamental elements of this type of problem: objective function, constraints and decision variables vector. For the FAME production objective function, the equality constraints are constituted of the fundamental equipment equations: material and energy balances, thermodynamic relations, sum of composition equations and kinetic expressions. The inequality constraints are composed of biodiesel purity specifications according to ASTM D6751, and glycerol concentration minimum specifications required. The decision variables vector is conformed of all of those taking part of the model and are not fixed and those that work as freedom degrees such as the water flow of the washing tower, the molar reflux ratio and the distillate rate of the biodiesel purification tower. On the other hand, the economic profit objective function comprehends two elements: i) investment capital costs, and ii) manufacture costs [5]. Equality and inequality constraints are the same of those in the operative function, but the variables that work as freedom degrees are the reaction operative pressure and the use of ethanol as raw material.

As resolution strategy for the operative optimization problem, the optimization rutine of Aspen Plus® modeling program was used, which uses Successive Quadratic Programming (SQP) Algorithm to solve the problem, sequentially looking for local optimal of equipment until it reaches a previously determined general convergence criteria [6].

The importance of this work is essentially its direct contribution to process engineering taking in notice that conditions and processes of biodiesel production are widely reported in literature, but there is no actual comparison between theoretical data and industrial reality. The final result of this study consists of establishing general guidelines to producers and future investors in this field, besides of a detailed economic evaluation of the optimized plant. As a general result, a validated simulation under Colombian industrial conditions, the optimal operative conditions and associated costs to the optimal configuration is presented. This conclusion pretends to collaborate and to continue with current researches and investigation made in the country referring to biodiesel production from palm oil. The impact of this work is ratified considering that Colombia is the fifth worldwide oil producer and the first in Latin America [7].

References

[1] Darnoko, D. & Cheryan, M (2000). Kinetics of Palm Oil Transesterification in a Batch Reactor. Journal of the American Oil Chemists' Society, 77, (12), 1263-1267. Recuperado el 21 de agosto de 2008 de la base de datos SpringerLink.

si3ea/documentos/documentacion/Biodiesel/Produccion_Biodiesel.pdf

[2] Zhang, Y., Dubé, McLean, D., & Kates, M (2003). Biodiesel production from waste cooking oil :1. Process design and technological assessment. Bioresource Technology, 89, (1), 1-16. Recuperado el 21 de agosto de 2008 de la base de datos ScienceDirect.

[3] Velosa, Federico. Simulación y optimización del proceso utilizado en Colombia para la producción de biodiesel a partir de aceite de palma. Bogotá, 2007, pp. 1-47. Tesis (Ingeniería Químico). Universidad de los Andes. Facultad de Ingeniería. Departamento de Ingeniería Química.

[4] CORPODIB: Corporación para el Desarrollo Industrial de la Biotecnología y Producción Limpia Recuperado el 17 de Octubre de 2008 de http://www.corpodib.com.co

[5] Turton, Richard. Analysis, Synthesis and Design of Chemical Process, 2ed. Prince Hall International, 2003.

[6] Fletcher, R. Practical Methods of Optimization, 1987. John Wiley & Sons, 2da edición, 436 pág.

[7] Federación nacional de productores de palma. Colombia. Recuperado el 13 de Octubre de 2008 de http://www.fedepalma.org.

See more of this Session: Modeling and Optimization for Manufacturing and Process Development
See more of this Group/Topical: Process Development Division

SOFTWARE INTEGRATION: EXCEL AND CHEMCAD

Flávio Waltz Moreira e Silva, Igor Ribeiro de Souza, Rafael Bertges Silva de Carvalho, Valter César de Souza,
Chemtech – A Siemens Company/Special Projects Division (SPD), Rio de Janeiro, RJ, - Brazil

Abstract

Simulations that involve a great amount of cases are becoming each time more frequent in optimization projects and sensitivity analysis. The current commercial software offers internal routines that solve some of the necessities of evaluation of situations as these, however useful mathematical treatment platforms does not communicated directly with process simulation software. The aim of the present work is to present a methodology where an integrated and synergic environment explores different simulation, optimization, visual and mathematical treatment environments advantages. Process simulator software CHEMCAD 6.0, form Chemstations, and the electronic spread sheet Microsoft Excel 2003 had been used, both software are available commercially.

Introduction

Many process simulation tools currently available in the market are used in different activities of different ways in the great majority of the applications. The most common softwares offers, in its majority, internal routines of sensitivity analysis and parameters optimization that covers some of the necessities of evaluation of situations that involve a large number of scenarios and/or changeable parameters.

On the other hand, the most powerful mathematical tools of treatment and numerical optimization are implemented in specific softwares in independent platforms and, in many cases, without any direct communication with process simulation softwares. Examples where it is necessary the accomplishment of a great number of simulation are easily imagined, however they can be not cogitated in an engineering project given the unfeasibility of simulations in an independent way for the Simulator of Processes.

The aim of the present work is to present the methodology developed in an integrated and synergic environment that explores the advantages of different simulation, optimization, mathematical and visual data treatment environment. It had been used the American process simulator software CHEMCAD in its version 6.0, and the electronic spread sheet Microsoft Excel 2003, both ones available commercially.

As example for the use of the methodology, a simple topology of an industrial plant of mixture of different combustible gases was modeled in the software, having as a goal of the simulations the evaluation of the properties of the outlet process mixture. It had been analyzed some operational parameters as volumetric fuel composition, fuel ratio and operational pressure and temperature. These parameters evaluation resulted in a total of approximately 90,000 simulations.

The great volume of data and the value of the evaluated parameters can be motivated by logistic and economic interests, such as properties specification and operational conditions of the final product at the delivery point.

Methodology

The main pillar for the implementation of the explored methodology was the use of a communication between programs tool, the DataMap. Such tool allows to make a direct connection (without a programming code) between an CHEMCAD simulation archive and a specific Excel spread sheet, through the allocation of cells of a spread sheet to physical properties generated by or defined in streams and/or unit operations.

The used methodology made use of three files:

- CHEMCAD simulation file;
- Spread sheet of communication;
- Storage spread sheet.

Each one of the used files has a specific purpose; however the key for the series simulation accomplishment possibility is the simultaneous and connected use of these files. A description of the purpose of each one of the files is described below:

CHEMCAD simulation file

It is in this file where the simulation topology is built. The complex calculations are responsibility of this file (thermodynamic properties, etc.) in accordance with the used topology.

Spread sheet of communication

It is this spread sheet that is indicated in the DataMap tool, that is, is in it where the cells are allocated to simulation streams and unit operations for the data insertion (before of the simulation) and/or withdrawal (after the simulation). It is this spread sheet also that contains the responsible codification for all information flow managing (entered, exit and treatment).

The programming code of the responsible routine for the simulations execution was developed in Visual Basic (VB) language and implemented in programming modules created in the its edition environment, available in Excel. The management of the information by the routine follows the following stages:

Dynamic mapping of communication spreadsheet - Attribution of the convenient input values to the mapped cells

Start of the simulation - Command of simulation initialization. The simulation runs using the values defined in the spread sheet and mapped by DataMap.

Data treatment - Mathematical treatment of the results calculated in the simulation, in order to get the desired values.

Results allocation - Transference of the calculated properties to the spread sheet of storage. The data are manipulated and stored in order to facilitate the late access and possible post-treatment.

Figure 1. Graphical representation of the information flow in accordance with the methodology described

Storage spread sheet

This spread sheet is responsible for the final simulations results storage after the VB routine treatment. It is to it where the results are destined after the ending of each one of the simulations.

When a simulation that uses the DataMap tool is executed in CHEMCAD, the software loads in memory, before initiating its calculations, the spread sheet which the DataMap is tied, and unloads the results at the end of the process.

The volume of data and consequently the file size where the information are stored grows in accordance with the course of the simulations, and in case that the spread sheet tied to the DataMap is the same one that storage the results, the simulations tends to be slower, as a consequence of the time expended in the loading and rescue of the spread sheet, which is proportional to its size. The use of an exclusive spread sheet for the storage of the data is necessary so that the increasing size of the file does not affect the time expended in each simulation.

Case Study

As illustration of use of the methodology proposal it was simulated a simplified PGL, Natural Gas (NG) and Air mixture unit, as presented in Figure 2. The equation of state (EOS) and the enthalpy model used was Soave/Redlich/Kwong (SRK).

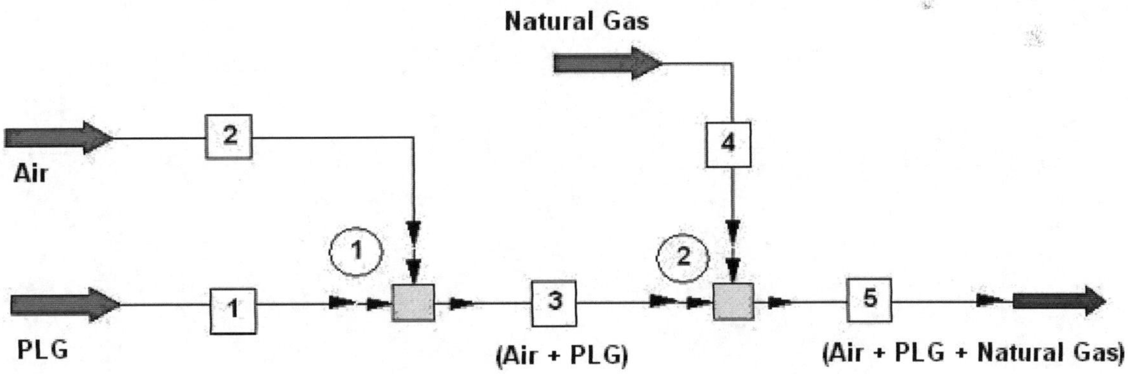

Figure 2. Topology of the simulation used as example

The simulations execution used the followed parameter variation:

- 2 Natural Gas compositions (11 molar fractions per composition);
- 10 PLG compositions (11 molar fractions per composition);
- 5 Volumetric ratios Air/PLG;

- 10 Volumetric ratios (Air+PLG)/Natural Gas;
- 15 Natural gas temperatures;
- 6 Natural Gas Pressures.

The combination of the previous parameters resulted in a total of 90.000 simulations.

In order to evaluate some specifics interest properties, some results was mathematical treated. The database was filled with the following properties:

- Temperature
- Composition (17 molar fractions per composition);
- Superior and inferior Wobbe Indexes;
- Inferior and superior inflammability limits;
- Inflammabilities components volumetric fraction.

Results

The 90,000 simulations had been carried through in 25 hours 13 minutes and 44 seconds (90,824 seconds), resulting in an average rate of 0,991 simulations/second. Such result was gotten using a computer with a 1GB memory RAM and Intel Pentium D 3.00GHz processor. The software had been used in the Windows XP operational system.

The profile of the simulation rate along the course of the simulations is shown in the Figure 3.

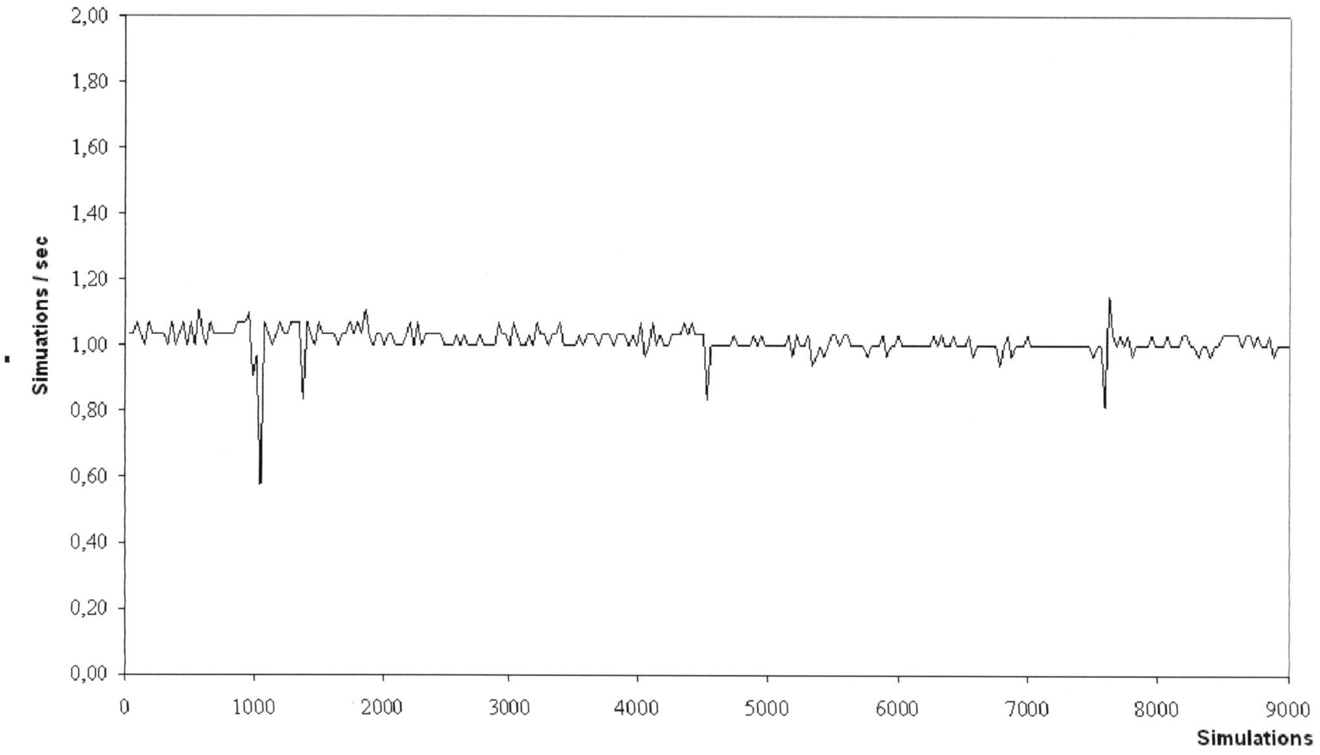

Figure 3. Simulation rate behavior

Conclusions

The open architectures to the programming and the data exchange between Excel and CHEMCAD software had been explored aiming to make possible the integration of two environments. The communication between software revealed sufficiently efficient, making possible and encouraging the joint and complementary use of the available tools in each one of the platforms.

The resources explored in the scope of this work was limited to present a small sample of the great amount of possibilities of the methodology use in the accomplishment of great numbers of simulation in a short period of time. The efficiency obtained in this work can be certainly applied in other analysis that needs a process simulator to generate a database for secondary calculations easily made in other platform software.

References

1. Crowl, D. A.; Louvar, J. F. (2002), "Chemical Process Safety – Fundamentals with Applications", *Ed. Prentice Hall.*
2. Soave, G. (1972), "*Equilibrium constants from a modified Redlich-Kwong equation of state*", Chem. Eng. Sci., vol. 27, pp. 1197-1203.

Dynamic Modeling and Control of the Downstream Units of a Hybrid Power/Chemical Plant

Patrick J. Robinson, Chemical Engineering, Lehigh University, Bethlehem, PA

William L. Luyben, Chemical Engineering, Lehigh University, Bethlehem, PA

Coal gasification is immerging as the premier unit operation in the energy and chemical industries. Gasification units in an electric power generation plant produce a fuel gas for driving combustion turbines. Gasification units in a chemical plant generate synthesis gas, which can be used to produce a wide spectrum of chemical products. In the future, plants are predicted to be a hybrid between power and chemical with the ability to handle unavoidable swings in both power demand and feed at high efficiency.

This paper presents an extension of the dynamic gasifier model, which uses a large hydrocarbon as coal assuming a 1:1 hydrogen to carbon ratio. The downstream units have been modeled in the widely used process simulator Aspen Dynamics. These dynamic simulations are essential for the development of stable and agile plantwide control structures of this hybrid power/chemical plant.

The downstream units of a typical gasifier facility producing electricity include purification of the syngas and shifting of syngas to hydrogen and carbon dioxide. Due to the high pressure of the system, hydrogen sulfide is removed by means of physical absorption. Selexol (a mixture of the dimethyl ethers of polyethylene glycol) is used for a product purity of less than 10 ppm H2S. Two water gas shift reactors then convert a total of 99% of carbon monoxide to hydrogen. Physical absorption of carbon dioxide with Selexol allows for essentially pure hydrogen to be fed into combustion turbines.

The coupling of a power plant with a chemical plant provides an additional control degree of freedom, fundamentally improving the controllability of the process. Swings in power demand can then be handled by diverting synthesis gas to a methanol plant for an easily-stored, hydrogen-consuming liquid product.

See more of this Session: Modeling and Optimization for Manufacturing and Process Development
See more of this Group/Topical: Process Development Division

Comparison of Methods for Estimation of Levels of Radiation Emitted by An Industrial Flare

Rafael Bertges Silva de Carvalho, Chemtech - A Siemens Company, Rio de Janeiro, BR
Antônio Dias Carrijo Neto, Chemtech - A Siemens Company, Rio de Janeiro, BR
Juliana Teixeira do Nascimento, Chemtech - A Siemens Company, Rio de Janeiro, BR
Fábio Sousa do Fundo, Chemtech - A Siemens Company, Rio de Janeiro, BR
Valter Cesar de Souza, Chemtech - A Siemens Company, Rio de Janeiro, BR
Flavio Waltz, Chemtech - A Siemens Company, Rio de Janeiro, BR

INTRODUCTION

The thermal radiation is the energy emitted by matter that is at a non-zero temperature. Although this phenomenon is commonly discussed for solids, it also occurs for liquids and gases. Regardless of the form of matter, the issue can be attributed to changes in the configuration of atoms or molecules that constitute it. In the heat transfer by radiation, the energy is carried by electromagnetic waves, the only form of heat transmitted without the intermediary of a medium material. In short, the radiation occurs more efficiently in the complete absence of matter, the vacuum.

Among the parameters that influence the exchange of heat by radiation, the following are the main: body material and the state of the emitting surface. The most common sources of thermal radiation are the sun and processes or reactions that produce flame. The heat radiation from the flames directly impacts on people and equipment, which can cause irreparable damage to human health, as well as structural damage, and the thermal fatigue, which is one of the main problems in metal structures of flares, sometimes requiring the use of shields heat to protect the structure.

The flare system has the function to perform the combustion of waste gases discarded from chemical processes in refineries and platforms. The burning must be done properly, ie, fulfilling the labor laws and not offering risks to people and facilities

In this study an analysis was performed between two different methods to determine the total amount of radiated heat, the one developed by Hottel[4] and the JFSH-Cook, which uses the fraction of radiated power, implemented in software PHAST®. The emitted radiation is used to simulate a platform, showing the levels of radiation in the geometry.

MATHEMATICAL MODELING

The calculation of the radiation levels emitted include in its initial stage obtaining/estimating the parameters and/or relevant conditions to the study, the output area of the flow, the conditions of emission of the mixture, the physical and chemical properties of the same, apart from parameters defining the flame generated, such as shape, length and area.

ENERGY RELEASED

The maximum amount of energy released (Heat release rate) for a generic flame can be obtained through the mass flow rate and the calorific value of the mixture burned. Thus, the rate of heat of combustion released can be obtained from the **Equation 1**.

$$Q = W \cdot \Delta H_{lv}$$

Equation 1. Heat release rate calculation

Where,

Q – Heat release rate(W)
W – Mass flow rate from the relief (kg/s)
ΔH_{lv} – Lower heating value (J/kg)

The **Table 1** shows the value of the rate of combustion heat released evaluated for a generic case.

Table 1. Heat release rate

Volumetric Flow Rate (m³/d)	Heat of Combustion (MJ/kg)	Heat Release Rate (MW)
690.000,00	48,08	321,76

However, it should be noted that only part of the rate of combustion heat released will be sent to the platform by radiation. The fraction of total heat radiated depends, among other parameters, the composition of the current and speed.

GEOMETRY

Usually represented by burners, directly affects the emissivity of the flame, and consequently the fraction of heat radiated. If the tentacles of the burner are more concentrated the tentacles, there will be the higher the area of release of waste gases (slower removal of waste gases) and the higher the emissivity of the flame. For this study we adopt an outlet area of 0.17 meters.

FLAME MODELING

The two methodologies used influence the modeling of the flame directly. The first is the estimation of the emissivity of the gases from the combustion and the second used the software PHAST ®, estimates the fraction radiated and directly results in the total energy radiated. Evaluation of some parameters of the flame generated by burning, as its emissivity, its shape and surface area is essential for assessing the radiation.

FLAME GEOMETRY

The geometric characteristics of the flame are modeled as a cone frustum with its smaller burners located next to the flare system. The geometric features of interest are the length of the flame and its major and minor bases, which enable the estimation of surface area and the total volume of the flame.

The **Table 2** shows the main operational characteristics (input data) and geometrics of the flame for the evaluated case.

Table 2. Geometric data for the flame

Volumetric Flow Rate (m^3/d)	Velocity (m/s)	Radius Minor/Major (m)	Flame Length (m)	Surface Area (m^2)	Volume (m^3)
690.000,00	46,81	0,90/5,54	24,75	608,43	945,69

ENERGY RADIATED BY FLAME

The heat release rate generated from the burning of the waste gas flow rate sent to the flare system, represents the maximum rate of energy released by the flare system. The rate of heat emitted by radiation is just a fraction of the rate of heat release rate. An estimate of the rate of heat radiation emitted by a flame is an extremely difficult process because the accurate determination of flame temperature and the presence of particles is a major challenge. The flame temperature is always lower than the adiabatic flame temperature and the contribution of the emission of particulates is difficult to be obtained due to uncertainty in their concentration. Flames with low presence of particles emit a small fraction of the energy released by combustion process [4].

The radiant energy emitted by the flame is just a result of emission of the gases and solids present in it. In this context, one can estimate the rate of energy released by the flame through the evaluation of the radiation emitted by combustion gases generated, assuming that there is a small fraction of particles present and the mixture is at the adiabatic flame temperature.

The gaseous products of combustion, water vapor and carbon dioxide, are responsible for emit in significant quantities in a flame. Thus, despite the complicated spectral and directional effects, a simplified procedure can be used to estimate the flow of thermal radiation emitted by a gas chromatography. The method was developed by Hottel [4] and involves the determination of emission of radiation by a gas chromatography temperature Tg. Thus, the heat flow can be represented by the **Equation 2**.

$$E_g = \varepsilon_g \cdot \sigma \cdot T_g^4$$

Equation 2. Radiated heat rate calculation

Where,

E_g – Heat rate radiated by the flame (W/m²);
ε_g – Gas emissivity (-);
σ – Stefan-Boltzman constant (W/m².K^4);
T_g – Gas temperature (K).

Thus, having the emissivity of the gas and adiabatic flame temperature, one can estimate the rate of release of heat radiated by the flame.

TEMPERATURE AND FLAME EMISSIVITY

Performing a conservative approach, the temperature of the flame was estimated as the adiabatic flame temperature. This consideration implies that the heat generated by combustion is converted into sensible heat of the gases generated and inert gases, so, the heat generated causes an increase in temperature of the gases involved. Thus, based on the composition of the gas relief, and considering a stoichiometric combustion, were determined: the fractions of water, carbon dioxide, nitrogen and trace gases after the reaction and their heat capacities and the temperature rise. The flame temperature obtained for the analyzed stream is 1834.20 ° C

Based on the flame temperature, one can estimate its emissivity, with the average beam length (L_ε), a parameter that correlates the dependency of the emissivity of the gas to the size and geometry of the flame, and its partial pressure [4]. Correlations for the average beam length can be easily found for simple geometries, like spheres and cylinders, however more complex geometries, like truncated cones, can have their L_ε estimated from the **Equation 3**.

$$L_\varepsilon = 3{,}4 \cdot \frac{V_g}{A_g}$$

Equation 3. Calculation of the average beam length for a given gas species

Where,

L_ε – Average beam length (ft);
V_g – Gas volume (ft³);
A_g – Surface area for the gas volume (ft²).

Thus, in possession of the average beam length, mole fraction of water and carbon dioxide, and the flame temperature, one can obtain the emissivity of these gases from the **Figure 1.** [4].

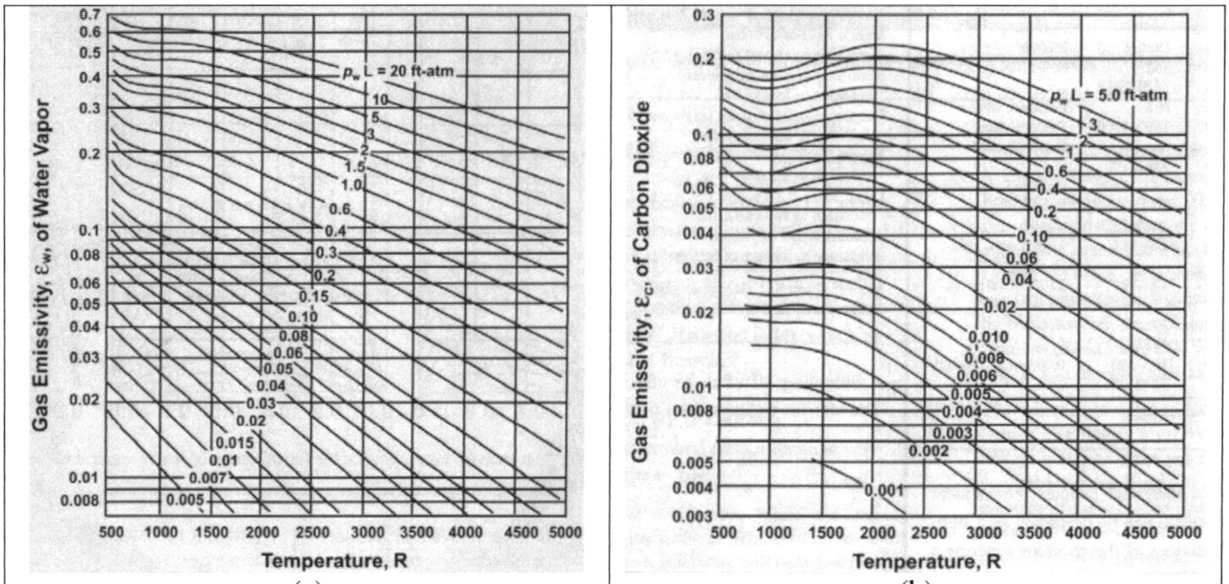

(a) (b)

Figure 1. Water vapor (a) and carbon dioxide (b) emissivities [4].

However, it should be noted that the emissivity values obtained in Figure 1. are related to the emissivities for pure gases, at 1 atm. Thus, for the case of water vapor and carbon dioxide are present, along with non-radiating gases, must be obtained the total emissivity considering a correction factor, as shown in Equation 4.

$$\varepsilon_g = \varepsilon_a + \varepsilon_c - \Delta\varepsilon$$

Equation 4. Total emissivity of the gaseous mixture calculation

Where,

ε_g – Gas emissivity (-);
ε_a – Water vapor emissivity (-);
ε_c – Carbon dioxide emissivity (-);
$\Delta\varepsilon$ – Correction factor for the gas emissivity (-);

The total emissivity factor correction of the gas can be obtained from the **Figure 2**.

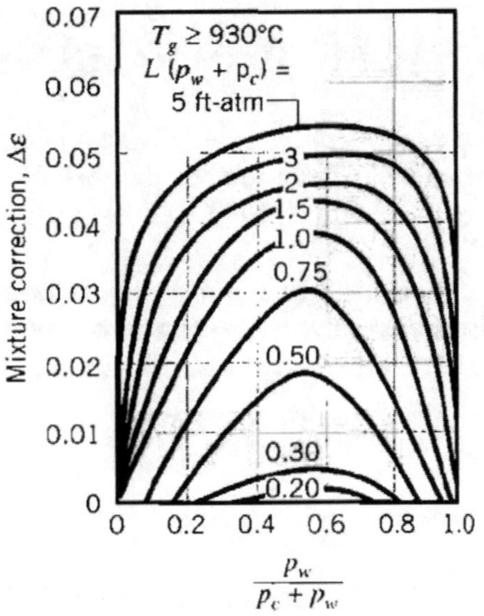

Figure 2. Total emissivity correction factor [4]

Thus, one can estimate the emissivity of the flame generated by the flare system. The **Table 3** shows the results obtained for these emissivities.

Table 3. Emissivity calculations results

Lε (ft)	**Lε.pᵢ (ft.atm)**			**pa/(pc+pa)**	**Emissivity**			
	H_2O	CO_2	H_2O+CO_2		H_2O	CO_2	Correction	**Total (gas)**
17,34	3,12	1,76	4,87	0,64	0,185	0,100	0,053	**0,232**

From the emissivity shown in **Table 3**, the calculated area of the flame and the **Equation 2**, one can determine the rate of heat radiated by the flame. Moreover, one can estimate the fraction of heat radiated. The **Table 4** shows the obtained values.

Table 4. Results for the Heat and Radiated Fraction

Emissivity	0,232
Rate of Heat Radiated(MW)	1.578
Radiated fraction	0,49

THE JFSH-Cook MODEL

Using the PHAST® software the levels of radiated energy can to be evaluated. The software uses the JFSH-Cook model, which is the Chamberlain Model modified. The model calculates the surface emissive power, which is the heat flux due to heat radiation across a flame's surface area.

The value obtained with that methodology is lesser that before found. This fact is a consequence of approximations and simplifications used in the evaluated of the flame temperature and in the emissivity of resultant gases. The **Table 5** shows the results obtained in both cases.

Table 5. Comparative between the methods

Method	**Radiated heat (MW)**	**Radiated fraction**
Hottel	1,578.00	0.49
JFSH-Cook	1,211.00	0.38

COMPUTATIONAL FLUID DYNAMICS SIMULATION

The results obtained can be use to simulate the physical phenomena of flare flame in an oil & gas platform, which provide evaluate the radiation levels near of equipments and operators. Both results can be used for that, however the JFSH-Cook is used because is a less conservative methodology.

In this step the PHOENICS® is used to evaluate the radiation levels. For this test, a general model of platforms is used. The **Figure 3** shows the contour of radiation levels between 14981 W/m² and 788 W/m², limits for people presence provided by API 521 [3].

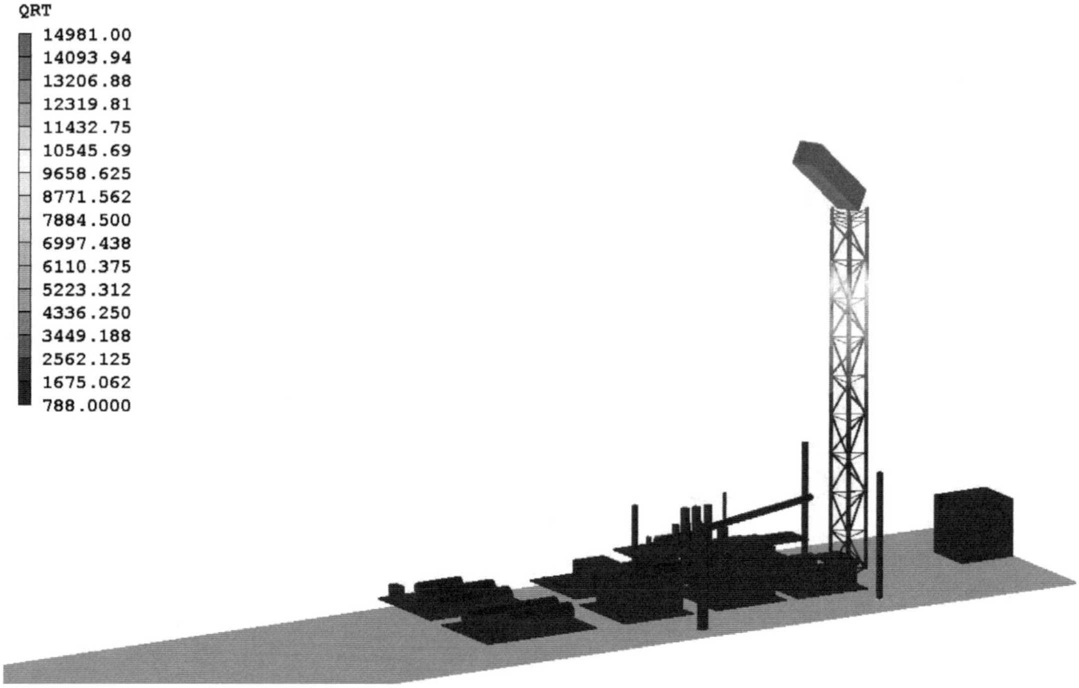

Figure 3. Radiation level

The **Figure 4** shows an iso-surface of 788 W/m², which is the surface where the radiation level is equal.

Figure 4. Iso-surface of 788 W/m²

CONCLUSIONS

Considering the results obtained, the methodologies can be used for evaluate the level of radiated energy emitted by flare flame and the JFSH-Cook model is less conservative which Hottel model.

The simulation shows the radiation levels near to equipments and structures. Therefore, the radiation levels can be seeing on the structure and at places where people presence is possible.

REFERENCES

[1] CROWL, D. A.; LOUVAR, J. F. (2002), "Chemical Process Safety - Fundamentals with Applications". 2nd edition. Prentice Hall. United States.

[2] PERRY, R. H. Perry's (1997), "Chemical Engineers' Handbook", 7th edition. McGraw-Hill. United States.

[3] "Pressure-relieving and Depressuring Systems" - ANSI/API STANDARD 521, 5th edition.

[4] Charles E. Baukal, Jr. (2001), "The John Zink Combustion Handbook".

[5] David P. Dewitt, Frank P. Incropera, "Fundamentals of Heat and Mass Transfer" - 6th edition.

[6] ISO 13702 – "Petroleum and natural gas industries - Control and mitigation of fires and explosions on offshore production installations" - Requirements and guidelines – 1st edition 1999-03-15.

Overpressure Scenario Analysis in Chemical Plants

Claudia Martins Silva, Special Projects Division, Chemtech - a Siemens Company, Rio de Janeiro, Brazil

Process safety measures have been a great concern in most chemical and petrochemical industries that make use of pressurized systems. Numerous historical relates show that accidents involving such systems can be catastrophic, resulting in deaths and damage of valuable properties. The frequency and magnitude of such accidents have made official regulations and internal company standards become more stringent. Accurate design and appropriate operation and maintenance of pressurized systems and their protective devices are mandatory.

In spite of the rigorous specifications, pressurized vessels have the potential for rupture and leakage, and explosions still occur throughout the world. The most common cause of such accidents is the failure of a pressure relief valve.

Pressure relief devices are used to protect pressurized systems from exceeding the design pressure. The pressure may be imposed by an external or internal source. The protective devices include safety valves, relief valves and safety relief valves.

A safety valve is a spring-loaded pressure relief valve, which is designed to open to relieve excess pressure and to reclose to prevent further flow after normal conditions have been restored.

In this contribution, the results of analysis carried out on more than a thousand safety devices and pressurized equipment in plants of major chemical and petrochemical sites in the USA are summarized. The analysis consisted of a systematic evaluation of existing pressure relief valves, vents and rupture disks, including overpressure scenario identification and evaluation as well as sizing calculations.

The procedures used for analyzing the safety devices, individual pieces of equipment and system of equipment are presented. Global analyses including the overall system, for instance total and partial power failure, cooling water supply failure, air instrument failure, etc, are also conducted. The approach for sizing a header/flare system is also presented.

Tavares, S.S.M., Pardal, J. M., Menezes, L., Menezes, C.A.B., D'Ávila, C., Failure Analysis of PSV springs of 17-4H stainless steel, Engineering Failure Analysis, 16 (2009), 1757-1764.

Manu, C. C., Birk, A. M., Kim, I. Y., Stress Rupture Predictions of Pressure Vessels Exposed to Fully Engulfing and Local Impingement Accidental Fire Heat Loads, Engineering Failure Analysis, 16 (2009), 1141-1152.

Goyal, R. K., Al-Ansari, E. G., Impact of Emergency Shutdown Devices on Relief System Sizing and Design, Journal of Loss Prevention in the Process Industries, 22 (2009), 35-44.

Koper, O., Westphal, F., Database-Supported Documentation and Verification of Pressure Relief Device Design in Chemical Plants, Journal of Loss Prevention in the Process Industries, 16 (2003), 73-79.

Malek, M. A., Pressure Relief Devices: ASME and API Code Simplified, McGraw-Hill, 2006.

See more of this Session: Modeling and Optimization for Manufacturing and Process Development
See more of this Group/Topical: Process Development Division

CFD MODELING OF FIRE SPREAD IN AN OFFSHORE PLATFORM

Renata Barros Pimentel Machado[1]
Juliana Teixeira do Nascimento[1]
Fábio Souza do Fundo[1]
Valter Cesar de Souza[1]

[1] *Chemtech Serviços de Engenharia e Software Ltda.*
Rio de Janeiro – Brasil.

Resume

This work presents the methodology used to model a fire spread on an offshore platform. CHEMTECH made this study to recommend the usage of passive protections on the structural elements on the platform. A computational fluid dynamics (CFD) program was applied to study the temperature range in the structural elements of the platform in case of a fire outburst. Initially, the results of a PRA (Preliminary Risk Analysis) made at the platform were used, and the principal data were input on PHAST® (Process Hazard Analysis Software Tool) - a risk analysis software. This analysis consisted in an initial estimate of the shape and dimension of the fire. Then, those results have been inputted on PHOENICS (Parabolic and Hyperbolic or Elliptical Numerical Integration Code Series) to simulate the impact of the fire on the structural elements of this platform. Four steps were followed: creation and simplification of the platform's geometry, mesh generation, input of boundary conditions and the simulation, each of them with its particular difficulty. CHEMTECH reached very goods results with this study and could recommend the usage of passive protection in the critical structural elements of platform.

1. Introduction

CHEMTECH developed a fire propagation study using computational fluid dynamics (CFD) as calculus and analysis tool. The software used was PHOENICS (*Parabolic and Hyperbolic or Elliptic Numerical Integration Code Series*), in version 2007, developed by the British enterprise CHAM *(Concentration, Heat and Momentum Limited*). With this software it was possible to simulate the physics processes numerically related with fire propagation in platforms, so as the temperature profiles along the structure could be obtained.

To model the fire originated from huge leaks the software PHAST® (*Process Hazard Analysis Software Tool*), from DNV, was used in its version 6.51. PHAST® is a risk analysis software used for project development and operation simulation. Jet fire and pool fire geometries were obtained and adapted for PHOENICS in which the temperature profiles on platform structures were analyzed. Based on obtained temperature fields the necessity of passive protection on structures with imminent risk was analyzed.

This article has the objective to present the methodology used by CHEMTECH to model a fire spread in an offshore platform. In this way, the stages involved in all simulation processes will be presented, with emphasis on each critical item of them.

2. Simulation Stages

2.1. Platform Geometry

The geometry developed for fire propagation study was made based on arrangement draws. The difficulties with the platform geometry were basically related to the simplification of the original one, since it had many details that would increase substantially the computational time during a simulation. The simplification of this geometry should avoid big changes that could interfere on the simulation results.

This process involved the deletion of beams and secondary structures, small objects and any geometry inside closed modules (rooms, accommodations) that would not have a leak.

Figure 1 makes possible a perspective visualization of the geometry used in simulations.

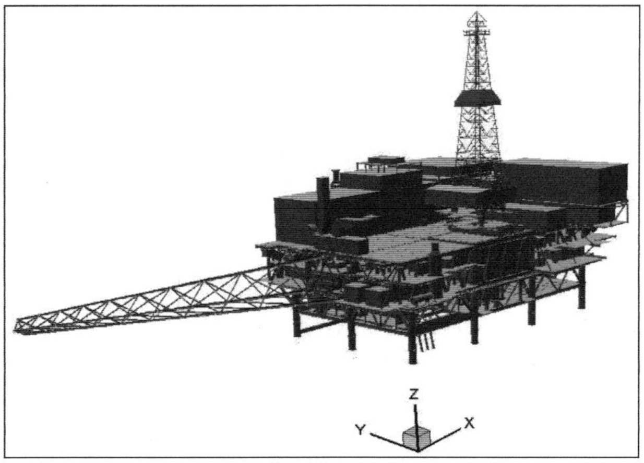

Figure 1: Platform geometry in PHOENICS.

Source: Own elaboration.

2.2. Flame Geometry

Studies of flame propagation are characterized by the simulation of large leaks originated by gas or oil on platforms. These possible leaks were analyzed during a PRA (preliminary risk analysis) made during other stages.

As mentioned before, the flame geometry was obtained by simulations in PHAST®, using a specific model according to the formed fire.

In this work the consequences generated by jet fire for gas leaks were analyzed. Jet fire occurs when the flame is developed by a combustible jet, released from a pressurized recipient. The software PHAST® has two models available for jet fire calculations: Shell and API 521. In this study the Shell model was used, because it foresees a geometry flame similar to a cone stem.

When an oil leak happens the pool fire occurrence is more probable. This kind of fire releases a lower energy than a jet fire, but its duration is in general longer. The geometry suggested by PHAST® was a cylindrical object, opaque and inclined by wind direction.

For the fire modeling the outputs of PHAST were adapted on PHOENICS, to obtain a better computational time and convergence. Jet fires were modeled as parallelepipeds with same length and area of the original geometry of PHAST and pool fires were modeled as cylinders with wind inclination as described in PHAST. In both cases the value of radiation obtained in PHAST was adopted. This fire geometry adaptation leads to a better computational result during simulation, making the problem convergence easier.

Flame directions for jet fires were made in accordance with leak directions, to the interior of the platform and sustentation pillars, establishing a conservative approach.

Figure 2 shows the model and position results of a flame for a scenario of Jet fire and a pool fire.

Figure 2: Flame model in platform – Jet Fire / Pool Fire.

Source: Own elaboration.

2.3. Mesh and Domain

For a CFD study, it's necessary to define the domain and the mesh for the development of the simulation. It was used a minimal domain, with the same size of the geometry, but with wind velocities already simulated and with convergence. With this methodology it was possible to reduce the cell, generating a mesh with closely 400.000 volume elements, reducing computational time. In the flame region refine in the mesh was made so that the temperature profile could be better observed.

2.4. Temporal Dependence and Contour Conditions

Simulations were developed in steady state for all cases in the study, with contour conditions like wind velocity, leak flow, and results of dimension and radiation intensity obtained by PHAST®.

Eight main wind directions were used in simulations, so as wind calms. For each wind direction the average velocity in the range of the more frequently ones was considered. For leak directions only the ones directed to interior of the platform and sustentations pillars were considered.

3. Conclusion

With the results obtained during this study, it was possible to observe structures that reached temperatures over than 400 ^{0}C. For these ones CHEMTECH recommend the usage of passive protection to prevent collapse in the critical structural elements of platform in case of a fire.

References

[1] PHAST Training Course Manual

[2] CROWL, D. A.; LOUVAR, J. F. Chemical Process Safety – Fundamentals with Applications. 2ª Edition. Editora Prentice Hall. USA, 2002.

[3] COTE, A. E. Fire Protection Handbook – Volume II. 19ª Edição. Editora National Fire Protection Association. Estados Unidos, 2003.

[4] PERRY, R. H. Perry's Chemical Engineers' Handbook. 7ª Edição. Editora McGraw-Hill. Estados Unidos, 1997.

[5] MENDES, M. F., Uma Metodologia para Análise Computacional de Incêndios em Instalações Offshore, Rio de Janeiro, 1996.

[6] The Netherlands Organisation of Applied Scientific Research, Methods for the determination of possible damage, Voorburg, 1ª ed, 1992. (TNO Green-Book CPR 16 E).

A New Approach for Facility Siting by Mapping Risks On a Plant Area Grid and Optimization

Seungho Jung, Chemical Engineering, Texas A&M University, College Station, TX

Dedy Ng, Artie McFerrin Department of Chemical Engineering, Mary Kay O'Connor Process Safety Center, College Station, TX

Carl D. Laird, Artie McFerrin Chemical Engineering Department, Texas A&M University, College Station, TX

Dr. M Sam Mannan, Artie McFerrin Department of Chemical Engineering, Mary Kay O'Connor Process Safety Center, College Station, TX

It is well known that for Facility Siting hazard scenarios shall be identified which can have significant effects on occupied buildings, the vulnerability of control rooms or other occupied buildings shall be determined at their location relative to the risk source, and the spacing of process units, and spacing between equipment and employees in occupied buildings be examined. Usually, such study is conducted for evaluating the location of process plant buildings which already exist. However, this problem can be approached from a different point of view which is to decide on locations of facilities in the initial design step instead of changing location later as this can help saving money.

This research focuses on achievement of optimal location of hazardous facilities and other process plant buildings by mapping risks on given land and optimization theory in order to minimize financial risk. Applying event tree analysis possible consequences such as VCE and BLEVE by flammable gas release at a hazardous facility, and their probability can be estimated. Failure frequency of containment gives input to estimating release probability and initial hole size. The given land area is divided in a square grid of a certain mesh size and each grid acquires a risk score given the type of consequence and type of exposed receptor/target relevant to the location. These risk scores such as the probability of certain structural damage are to be multiplied by the costs of potential facilities to be built on the grid finally giving us the financial risk.

Accompanying the suggested safety concept, the new model takes into account economics. The overall cost of placing a facility on a certain location is a function of piping cost to connect it to adjacent facilities, land cost, and financial risk. The proposed approach is an assignment problem that identifies attractive locations by minimizing overall cost. This approach gives the coordinates of each facility, and estimates the total piping length including management cost. The main purpose of this paper is to assist the selection of locations for process plant buildings and managing the risk.

See more of this Session: Applications of Optimization in Industrial Practice
See more of this Group/Topical: Process Development Division

Experimental Validation of An Approach to Real-Time Economic Optimisation

Bram Satriadarma, Chair of Process Dynamics and Operation, Berlin Institute of Technology, Berlin, Germany

Harvey Arellano-Garcia, Chair of Process Dynamics and Operation, Berlin Institute of Technology, Berlin, Germany

Gunther Wozny, Chair of Process Dynamics and Operation, Berlin Institute of Technology, Berlin, Germany

This work discusses the implementation and integration of Predictive Functional Control (PFC) with the Economic Optimisation for close loop plant-wide optimisation, on a pilot distillation column. Backx et al. proposed a multilayer control strategy for close loop plant-wide optimisation for "market driven mode" industries, which separated optimization and control layer for practical feasibility due to complexity of the full dynamic optimization of a plant. The consistency between model-based optimisation and unit-level control (MPC) is crucial for high performance in plant operation. Therefore, it is necessary to fully integrate the optimisation and control layer. Moreover, Tosukhowong et al. focused on optimizing the real time optimisation RTO (optimisation layer) in order to have a better "multilayer control strategy" performance. In contrast to this, in this work we focus on the multilayer control strategy by integrating PFC (Predictive Functional Control) in the control layer, because of the advantages from PFC and its successful implementation in industries [2,9]. In order to demonstrate and verify the efficiency of the developed approach, a multilayer control strategy for a high pressure distillation column is used as an example.

[1] Sloley, A.W. (2001): Effectively Control Column Pressure

[2] Richalet, J.(1993): Industrial application of model-based predictive control. Automatica, vol. 29, no. 5, pp. 1251-1274.

[3] Morari, M. and Zafiriou, E. (1989): Robust process control. Prentice Hall.

[4] Joe Qin, J., Badgwell, T.A. (2003): A Survey Of Industrial Model Predictive Control Technology. Controll Engineering Practice 11, 733 - 764.

[5] M. Lawrynczuk, P.M. Marusak and P. Tatjewski (2007): Efficient Model Predictive Control Integrated with Economic Optimisation. 2007 Mediterranean Conference on Control and Automation.

[6] M. Lawrynczuk, P.M. Marusak and P. Tatjewski (2007): An Integrated Structure with Economic Optimisation and Constrained Model Predictive Control: an Application to a MIMO pH Reactor. EURCON 2007.

[7] T. Tosukhowong and J.H. Lee (2004): Real-time Economic Optimization for an Integrated Plant via a Dynamic Optimization Scheme. Proceeding of the 2004 American Control Conference.

[8] T. Backx, O. Bosgra, W. Marquardt (2000): Integration of Model Predictive Control and Optimization of Processes. Proceeding of the IFAC Symposium "Advance Control of Chemical Processes", 249-260.

[9] Satriadarma, B., Arellano-Garcia, H. & Wozny, G. (2008) Implemetentation of Predictive Functional Control on a high pressure distillation column. AICHE 2008. Philadelphia, USA.

See more of this Session: Applications of Optimization in Industrial Practice
See more of this Group/Topical: Process Development Division

Novel Tool for Optimizing Petrochemical Derivatives Complex

AIChE Annual General Meeting. Nashville, TN.
Monday, November 9, 2009.

Atun Anggara,
Austin Crooks
and Osman Aboul-Nasr*

Fluor

Greenville, SC. USA

FLUOR®

Disclaimer

Fluor makes no representation or warranty, expressed or implied, as to the accuracy or completeness of the information contained herein and nothing contained herein is, or shall be relied upon as a promise or representation to be used in connection with any proxy, proxy statement, proxy soliciting materials, prospectus, Securities Registration Statement or similar document without the express written consent of Fluor, except as may be required by law.

FLUOR®

Fluor Corporation

- One of world's largest, publicly owned engineering, procurement, construction, and maintenance companies
- *Engineering News-Record (ENR)* magazine consistently ranks Fluor Corporation among the top three on its "The Top Design-Build Firms" and "The Top 100 Contractors by New Contracts" lists
- An international workforce of more than 30,000 employees worldwide
- Founded on a legacy of delivering world-class solutions for almost 100 years
- One of the world's safest contractors

FLUOR®

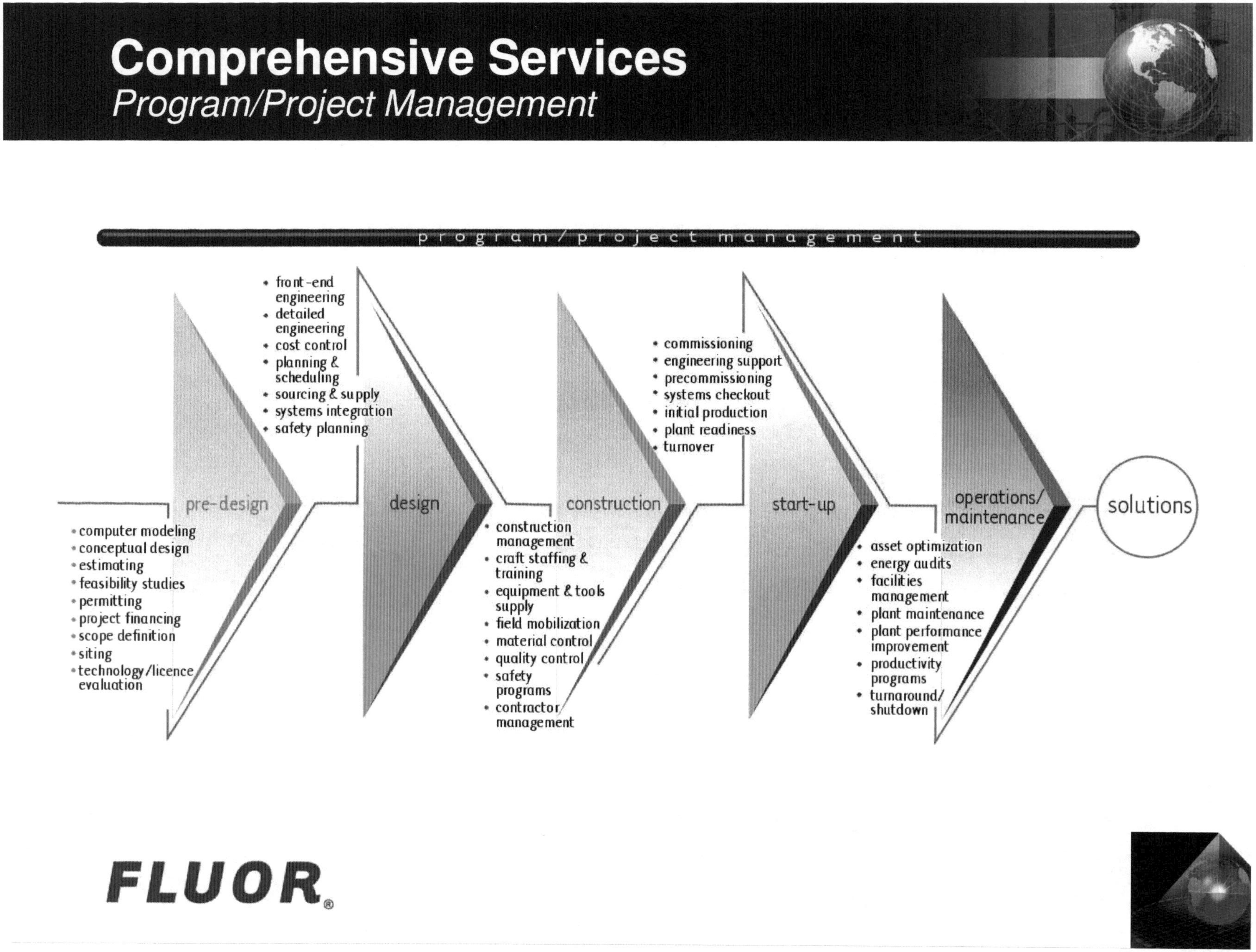
Comprehensive Services
Program/Project Management
program/project management
pre-design
computer modeling
conceptual design
estimating
feasibility studies
permitting
project financing
scope definition
siting
technology/licence evaluation
design
front-end engineering
detailed engineering
cost control
planning & scheduling
sourcing & supply
systems integration
safety planning
construction
construction management
craft staffing & training
equipment & tools supply
field mobilization
material control
quality control
safety programs
contractor management
start-up
commissioning
engineering support
precommissioning
systems checkout
initial production
plant readiness
turnover
operations/ maintenance
asset optimization
energy audits
facilities management
plant maintenance
plant performance improvement
productivity programs
turnaround/ shutdown
solutions
FLUOR®

Outline

- Master Planning
- Market Study
- Capability of the Software
- The Chemical Tree
- Example of Adding Value
- Conclusions

FLUOR®

Master Planning

- **Feedstock Available**
  - How much
  - Composition
- **What should we do with it?**
- **How much will cost?**
- **Is it economically feasible?**

FLUOR®

Market Study

- Target markets
- Potential products and capacities
- Price forecast
- Corporate strategy

FLUOR®

Software Capabilities

- **Cracker and downstream configurations**
- **25 Built-in cracker capabilities**
- **Over 400 product-licensor combination**
- **Class 5 TIC updated annually (USGC Basis)**
- ***Non-Confidential Information***
- **Utilities, direct labor and Working Capital**
- **Output**
  - TIC
  - IRR
  - Other basic information.

FLUOR®

Cracker Capabilities

INPUT

- **Feedstock**
  - Ethane through naphtha
  - Various severities
- **Built-in**
- **Fluor's Database**
- **Client's information**
- **Commercial software**

OUTPUT

- **Products**
  - Ethylene
  - Propylene
  - C4's (Recycle)
  - Pygas
  - Heavies
- **Utilities**
- **Labor**

FLUOR®

The Chemical Tree: Product and Technologies Available

(Full scale demo during session)

FLUOR®

Ethylene (Available Products)

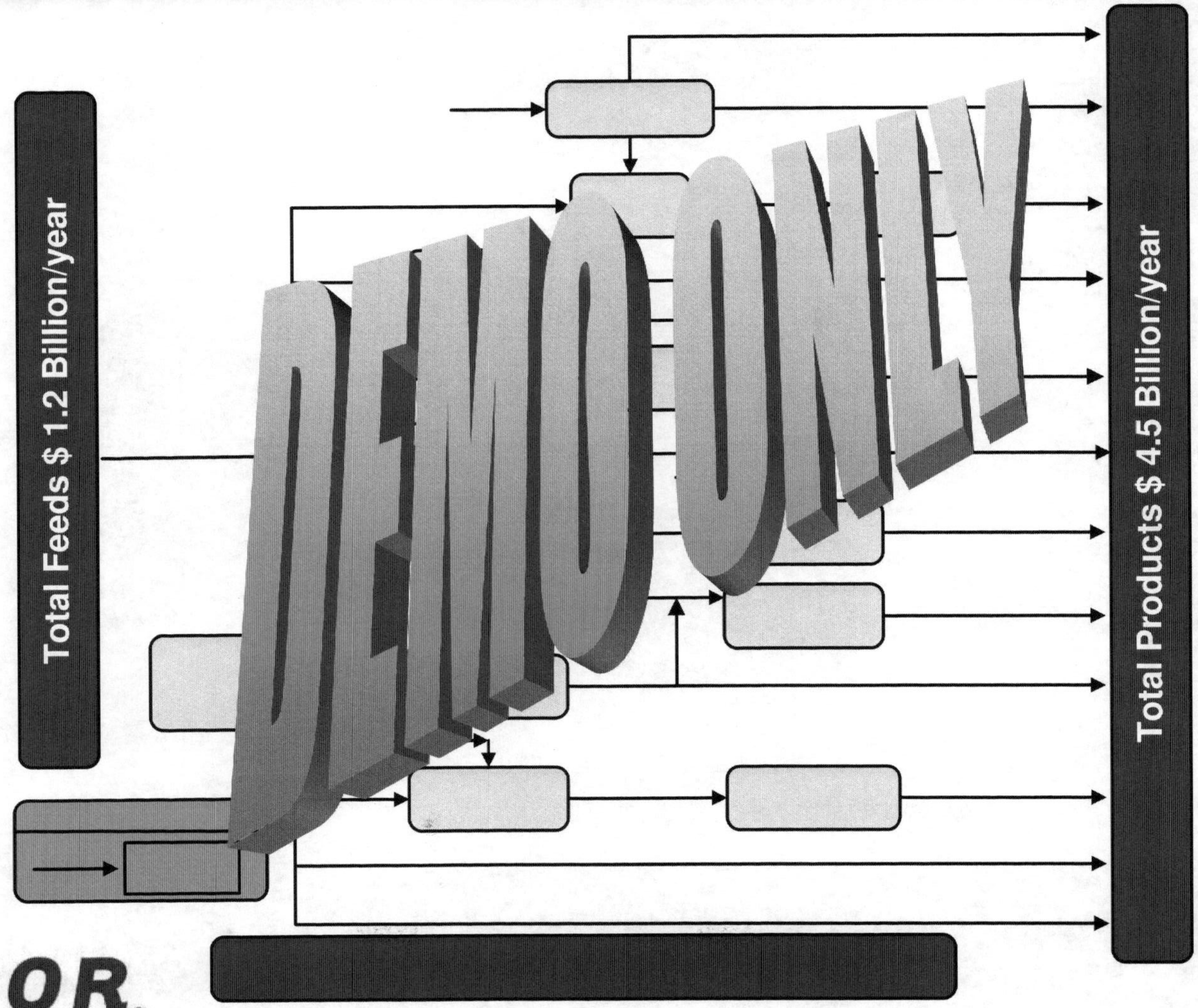

FLUOR®

Option 1

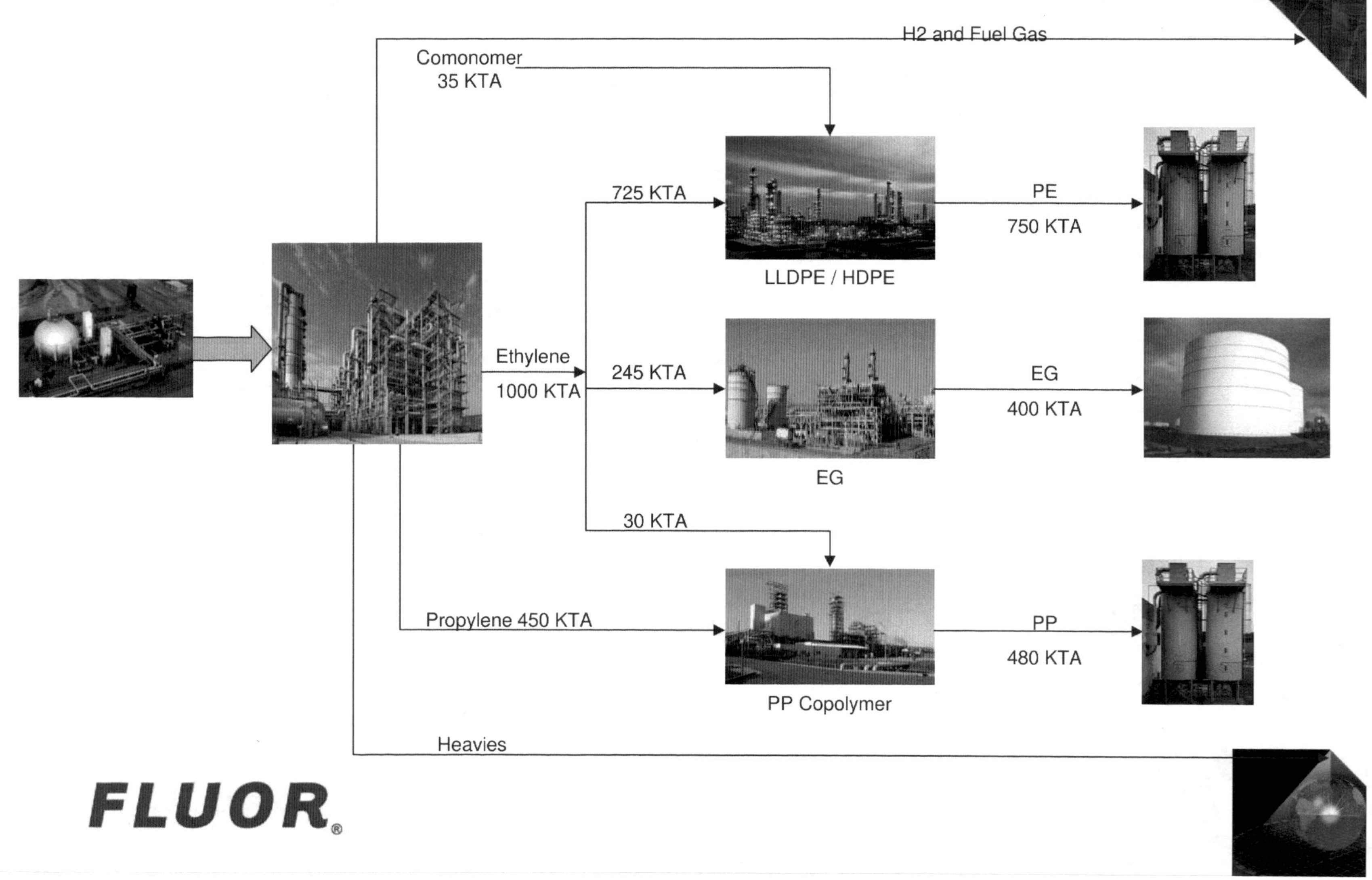
H2 and Fuel Gas
Comonomer
35 KTA
725 KTA
LLDPE / HDPE
PE
750 KTA
Ethylene
1000 KTA
245 KTA
EG
EG
400 KTA
30 KTA
Propylene 450 KTA
PP Copolymer
PP
480 KTA
Heavies
FLUOR

Option 2

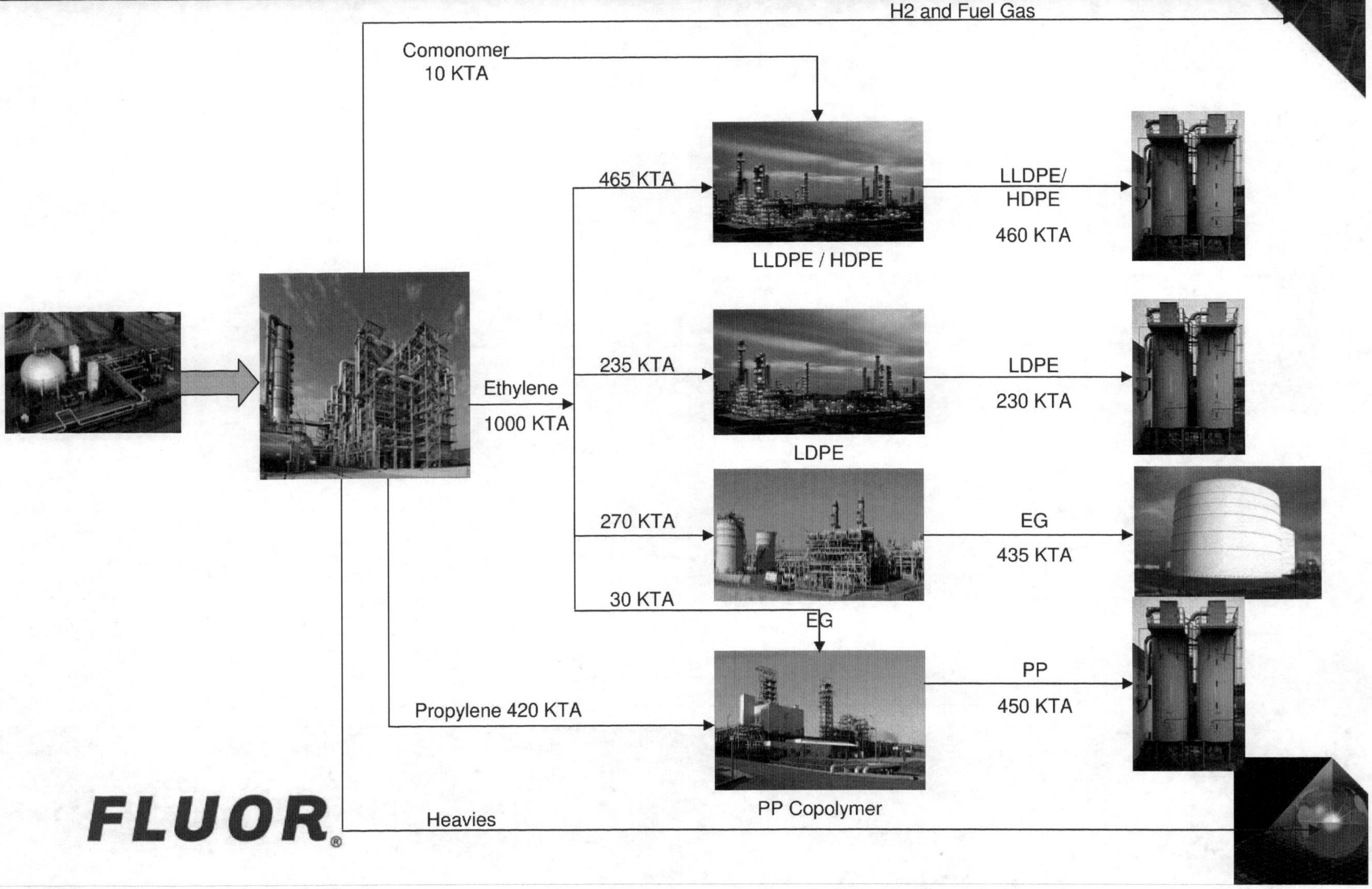

FLUOR®

Example 3

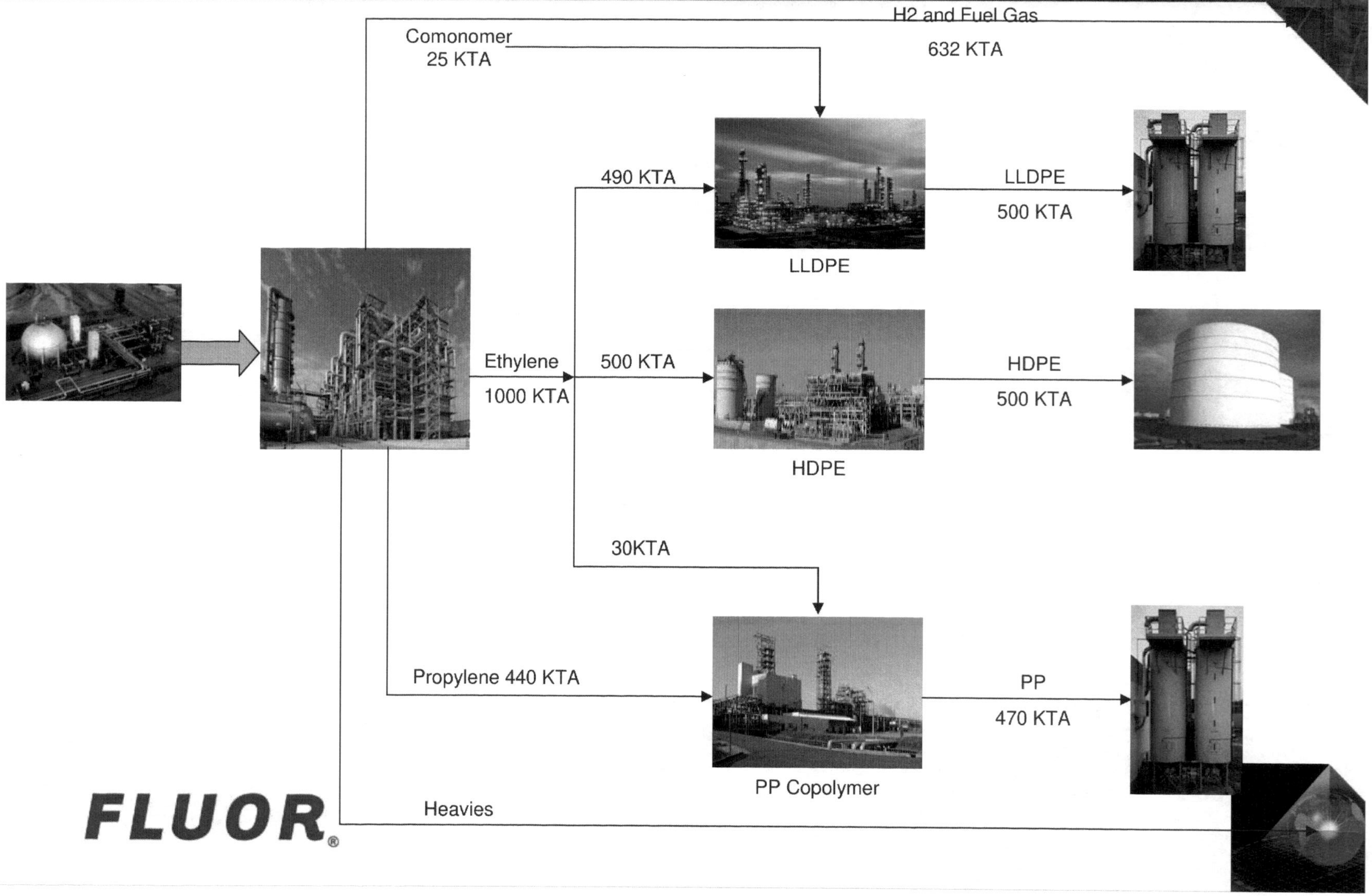

H2 and Fuel Gas
632 KTA
Comonomer
25 KTA
Ethylene
1000 KTA
490 KTA
500 KTA
30KTA
LLDPE
HDPE
PP Copolymer
LLDPE
500 KTA
HDPE
500 KTA
PP
470 KTA
Propylene 440 KTA
Heavies
FLUOR®

Demo During Session

- How the TIC can be developed
- How the economics are generated
- Options developed in minutes

FLUOR®

Conclusions

- Master Planning incorporates corporate strategy, technical and economic feasibility
- A tool for quick Class 5 estimates is available
- Options developed in minutes.
- Discard outrageous options.
- Out-of-the-box options can be tested quickly.
- Choose best options for further development.

FLUOR®

A Method for the Computation of Many Local Optima in Process Design Problems

M. Urselmann, S. Engell: Process Dynamics and Operations Group, Technische Universität Dortmund, Germany, e-mail: maren.urselmann@bci.tu-dortmund.de

Introduction

Previous work on the optimization of the design parameters of complex processes, especially of reactive distillation processes, has shown that in these problems, a large number of local optima exist [1], [2]. While some of these local optima are close to each other and are most likely caused by early termination of the nonlinear optimization algorithm, many of them are structurally different, i.e. they have different catalyst distributions on the trays and different feed locations. It is therefore of practical interest to identify most, if not all of these local optima in order to make design choices, e.g. by taking additional information into account, like information on the uncertainty of the assumptions made in the cost function, or heuristic knowledge on favorable designs.

Such design optimization problems are usually tackled by mixed-integer nonlinear programming (MINLP) techniques, e.g. DICOPT [3], where the optimization of the structural decisions, of the design parameters and of the operational degrees of freedom is performed simultaneously with the determination of the corresponding state variables. This leads to large optimization problems with several thousand variables and as many complex constraints that are implicitly given by chemical and thermodynamic equilibria and the reaction kinetics. Even without considering integer variables as e.g. the number of trays, the resulting nonlinear optimization problems (NLPs) can only be solved reliably after careful reformulation of the equations and scaling of the variables [4] and provide only one local optimum. The application of stochastic multi-start heuristics improves the probability of finding the global optimum, but then the numerical effort increases by typically two orders of magnitude [1], [2].

In this work, an evolutionary algorithm (EA) is used to generate initializations for a local search method, such that a large number of local optima is identified. The efficiency of the search is increased by the introduction of so-called tabu zones which exclude regions from the future search that are defined by the initial points of the local search and the corresponding local optima [5]. As it is too complex to handle the full search space that consists of the design variables and the state variables of the resulting process in the EA, the evolutionary search is confined to the space of design variables. The state variables that are needed for the evaluation of the cost function and of the constraints are computed by the same nonlinear solver that performs the local optimization. The combination of the EA and the NLP solver results in a memetic algorithm (MA) [6].

Case study

The considered reactive distillation process is the heterogeneously catalyzed and kinetically controlled synthesis of methyl-tertiary-butyl-ether (MTBE) from isobutene and methanol in the presence of n-butane at a pressure of 8 bar. The mixture exhibits three binary azeotropes. The aim is to produce MTBE with a purity of 99 % from feed streams of methanol (6.375 mol/s) and isobutene/n-butane (8.625 mol/s, 65.2 % isobutene according to [7]). The reboiler and the condenser of the distillation column are modeled as trays. It is assumed that fractions of both feed streams can enter the column on each tray including reboiler and condenser and that there can be a certain amount of catalyst on each tray of the column, restricted by the volume of a tray.

In the reactive distillation column, the chemical reaction is coupled with the distillation of the reaction mixture. This enables to overcome the chemical and thermodynamic limitations. In the conceptual design of reactive distillation columns, the operating parameters, the dimensions and the structure of the column are determined. The design is evaluated by its total annual profit that is composed of product revenues reduced by raw material cost, operating cost and annualized investment cost that depend nonlinearly on the design parameters. The main design variables are the locations of the feeds (initially the number of feed locations is not restricted), the distribution of the catalyst, the diameter of the column and the boil-up and reflux ratios. A design problem hence has 149 N+14 continuous degrees of freedom. Here, the number of trays (N) is assumed to be fixed. By using the formulation described in [4] empty trays are included, hence the number of trays is only an upper bound to the effective number of trays.

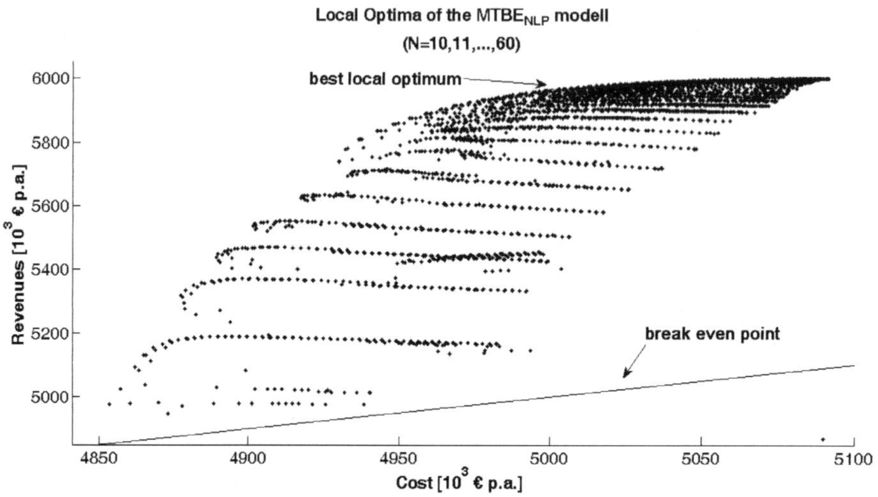

Figure 1: Local optima of the column design problem in the space of the cost (for N = 10, 11, ..., 60).

In Figure 1, the multi modality of the design problem at hand is shown. Each of the points displayed represents one of the local optima known so far for the column design problem with fixed numbers of trays, N = 10, 11, ..., 60 in the space of the cost and of the revenues. The plotted line indicates the threshold between profit and loss. It can be recognized that the quality of the local optima and the composition of the cost and of the revenues of solutions which give approximately the same profit are varying significantly. These variations reflect the differences of the design variables and the model variables at the local optima. The computation time for one local optimization started from a trivial starting point in the space of all variables depends on the number of trays N and varies between approximately 10sec for N = 10 trays and 6min for N = 60 trays. These times were measured on a PC with 2.66 GHz (Intel Xeon) and 3 GB of memory. For a more detailed description of the design problem at hand see [1] and [9].

The *optimization model* of the case study (in the following denoted by $MTBE_{NLP}$) contains all variables (model variables and design variables) as free decision variables. It consists of 1564 - 9314 algebraic equations (depending on the number of trays) and is formulated in the modeling language GAMS [10]. The *simulation model* comprises a subset of the equations and of the inequalities of the optimization model. The design variables were removed from the set of free variables. The equations and inequalities that restrict the feasible values of the design variables were also removed from the set of constraints. One of the constraints couples the diameter with the values of the catalyst amount and with some other model variables. To eliminate the problem that the equations in the model cannot be fulfilled for fixed design variables because of this constraint, the diameter and the related constraint were removed from the simulation model and are handled by the design optimization.

The optimization algorithm

The design optimization is tackled by a memetic algorithm (MA) that consists of an evolutionary strategy (ES) as described in [11] and a mathematical programming (MP) method. The ES generates new starting points for the MP solver (CONOPT [8]). The local problems are solved in a sequential manner. First, the state variables are computed for the initial point in the space of the design variables. Then the local optimization is performed simultaneously in the space of the design variables and the state variables. This leads to a significant reduction (>75 %) of the computation time needed for the local search. In Figure 2 the structure of the memetic algorithm is shown.

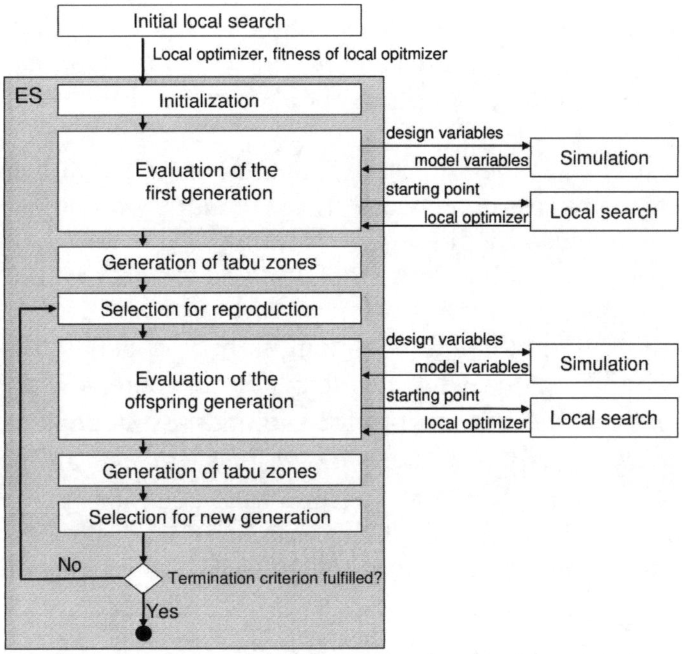

Figure 2: Structure of the memetic algorithm (MA)

At the beginning of each optimization, the MP solver is started from a trivial initial point, where all variables have the value 1. This first local search is necessary to find initial values for the model variables within the simulation model, because the initialization is crucial for the solution of the model equations. The values of the model variables at the first local optimum serve as initial values for the next simulation. During the optimization procedure all feasible points are saved. Before the simulation of any new point, the nearest of the feasible points found so far is determined. Then the values of its model variables serve as initial values for the simulation of the new point. The similarity of two points is measured by the Euclidean distance in the space of the design variables which are scaled to lie in the interval [0,…,2].

Subsequently the ES is started. The starting population of size μ is initialized randomly within the search space of the design variables, respecting the constraints that ensure that the sum of the feeds corresponds to the specifications. To evaluate the individuals of the initial population, the simulation model is solved for each individual. If the local solver cannot fulfill the model equations, it aborts at an infeasible point. An analysis revealed that the violation of the constraints at these points is only caused by the fixed reflux and boil-up ratio. Therefore these points can be used as starting points for the subsequent local search. The resulting points in the space of all variables are passed to the MP solver in order to perform a local search. According to the evolutionary model of Lamarck, the genes of the individuals are replaced by the values of the design variables of the corresponding local optimum.

For each local optimum, a region is defined (a *tabu zone*) to approximate the basin of attraction of this optimum. These regions are excluded from the subsequent search procedure of the EA.

To create the λ offspring, λ times an individual is chosen randomly from the population. Each of these individuals is mutated by a modified variant of the standard mutation operator described in [12]. The modification ensures that the individual remains feasible with respect to the constraints defined on the feed variables and that no offspring is created that lies within the tabu zones. A self-adapting step size introduced by [11] is applied. The initial step size is chosen with respect to the size of the feasible domain of the corresponding variable. A detailed description of the operators is given in [5].

Then the simulation is called for the resulting offspring individuals, and the local searches are started from the resulting values of the model variables. The genes of these offspring individuals are replaced by the design variables of the corresponding local optimum. For each of these local optima, a tabu zone is defined and the list of tabu zones is expanded.

If the termination criterion (here a maximum number of generations) is not fulfilled, the generation cycle is started again with the selection of parent individuals for the next generation. The standard selection procedure for evolutionary strategies is the so called (μ, κ, λ)-selection that chooses only the μ best individuals out of all parents and offspring individuals that do not exceed a maximal life-span of κ generations. The application of this selection procedure in combination with the replacement of the genes following Lamarck's approach caused a loss of diversity in the population of the ES. Therefore a new selection procedure was developed that ensures that no duplicates are chosen to survive whenever the population contains at least μ different individuals. In case of more than μ different individuals, μ randomly chosen individuals are selected. Each individual that corresponds to a local solution is stored immediately after its evaluation. Thus, none of the individuals that are replaced by this procedure gets lost.

The overall aim of the optimization is to find as many local optima as possible.

The tabu zones

In order to approximate the basin of attraction, tabu zones of different forms and sizes and with different locations in relation to the corresponding local optima were tested. Tabu zones with three different forms (rectangular, spherical and elliptical) were examined. The locations of the starting points and of the corresponding local optima define the locations of the tabu zones. All examined zones are symmetric around their centers. Their size is determined by the distance between the starting point and the corresponding local optimum. If the local optimum is taken as the center, the zones are called *symmetric*. If a point on the line between the starting point and the corresponding local optimum is chosen to be the center, then the zones are called asymmetric. The five different types of zones that were investigated are shown in Figure 3 schematically in a two-dimensional space.

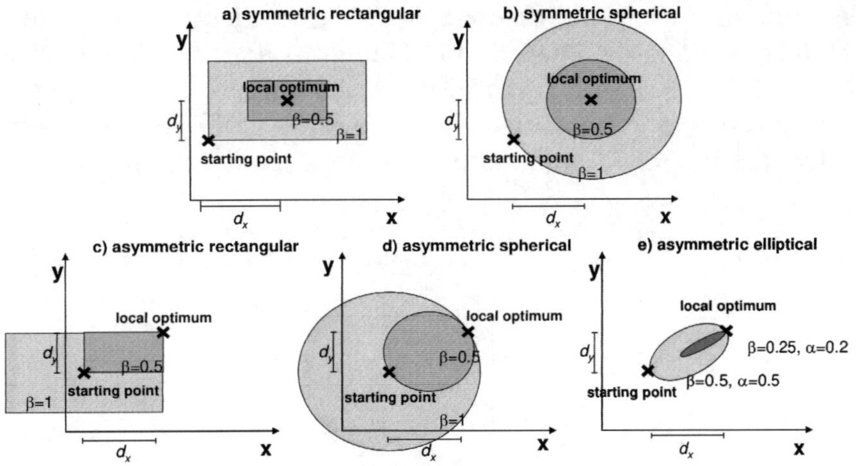

Figure 3: Schematic structure of the tabu zones

To vary the size of the tabu zones, a scaling factor β is defined. In case of the elliptical zones an additional factor α is defined to scale the second semi-axis. A scaling value of β=1 corresponds to the distance between the starting point and the local optimum. A scaling value of α=1 means that α equals β. Before the computation of the radius of the spherical tabu zones and the semi-axes of the elliptical tabu zones, the values of the variables were scaled to be in the interval [0,...,2].

Results

After some tuning experiments applied to three different model instances the parameters for the ES were fixed to the values that led to the best results. As a measurement for the quality of the results the number of local optima found within 50 generations of the MA and the corresponding success rates within 20 test runs were used. The population size was chosen to be μ=5, the size of the offspring population λ=10 and the elliptical tabu zones with scaling factors of β=0.4 and α=0.4 were chosen. In comparison to the application of the MA without any tabu zone the robustness of the algorithm could be improved significantly.

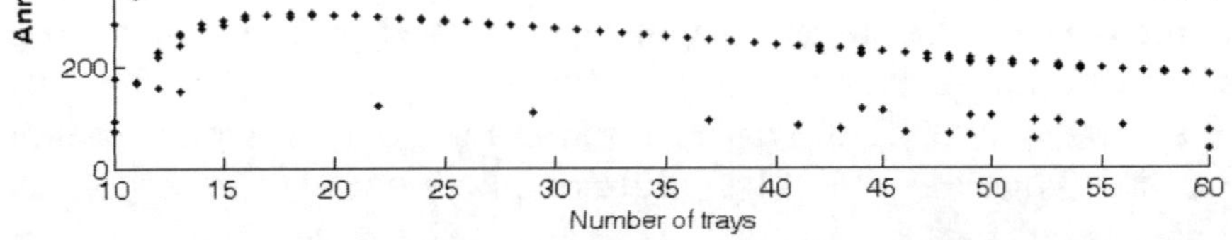

Figure 4: Local optima of the MTBE$_{NLP}$ model found by the MA in two runs, N = 10, ..., 60

For problem instances with a fixed number of trays, up to 214 local optima (for N = 59) could be located by the application of the new algorithm. By two runs for each model instance with N = 10, ..., 60 with a termination criterion of 100 generations, 5338 different local optima were found. The overall computation time needed for these runs were 20 days and 14 hours. All these optima are illustrated in Figure 4 ordered by the number of trays and the annualized profit. The memetic algorithm generates on the average one new local minimum for every 10 calls of the nonlinear solver (in case of the more complex models with N > 35 it generates one new local minimum for every 7 calls). In Figure 5 the best column design found is shown.

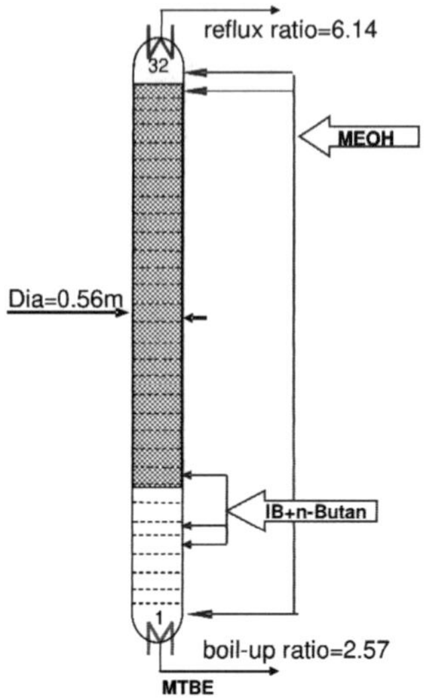

Figure 5: Best column design found by the MA.

It has N = 32 trays, three streams of each feed and a connected reaction zone formed by the trays 9 to 31. The reflux ratio is 6.14 and the boil-up ratio 2.57. It leads to an annual profit of 965 592€. 363 of the local optima located by the MA have objectives that are greater than 950 000€ and they correspond to instances with 25 up to 45 trays. The best 6 have objectives that are greater than 965 000€ with 31 to 34 trays.

Outlook

Many of the local optima found are unrealistic solutions because they correspond to a column design with many feed streams. In practice such a column would not be built. The approach will be extended by a restriction of the number of feeds to a maximum of 3 feed locations per stream. This extension introduces many discrete variables into the problem, because the number of feed streams and the location of the corresponding feed trays have to be optimized. These variables will be addressed by the EA, while the model will be kept continuous, because EAs do not need continuous and smooth functions to work properly whereas the MP solver does.

As an alternative, the cost function will be extended by the cost of the feed streams which leads to discontinuities when a stream becomes zero. Such functions can not be handled by NLP solvers. Therefore the EA will address the overall global optimization of the extended cost function while the NLP solver will not take the cost of the feeds into account.

References:

[1] S. Barkmann, G. Sand and S. Engell: Modellierungsansätze für die Design-Optimierung von reaktiven Rektifikationskolonnen, *Chemie Ingenieur Technik 80*, pp. 107–117, 2008.

[2] S. Barkmann, G. Sand and S. Engell: Optimisation-Based Design of Reactive Distillation Columns, European Congress of Chemical Engineering 6, pp. 435-436, 2007.

[3] J. R. Jackson, I. E. Grossmann: A disjunctive programming approach for the optimal design of reactive distillation columns, *Computers and Chemical Engineering 25*, pp. 1661-1673, 2001.

[4] G. Sand, S. Barkmann and S. Engell: Structuring of Reactive Distillation Columns for Non-Ideal Mixtures using MINLP-Techniques, European Symposium on Computer-Aided Chemical Engineering 14 (18), pp. 493-498, 2004.

[5] M. Urselmann, G. Sand, S. Engell: A memetic algorithm for global optimization in chemical process synthesis, Proc. Congress of Evolutionary Computation 2009

[6] P. Moscato: On evolution, search, optimization, genetic algorithms and martial arts: Towards memetic algorithms, California Institute of Technology, Pasadena, California, USA, Tech. Rep. Caltech Concurrent Computation Program, Report. 826, 1989.

[7] J. Stichlmair and Th. Frey: Mixed-Integer Nonlinear Programming Optimization of Reactive Distillation Processes, *Ind. Eng. Chem. Res. 40*, pp. 5987, 2001.

[8] A. S. Drud, CONOPT: A large-scale GRG code, *ORSA Journal on Computing* 6, pp. 207–216, 1992.

[9] G. Sand, S. Barkmann, P. Scherpian, M. Tylko, S. Engell and G. Schembecker: MINLP-Optimization in the Integration of Reaction and Separation Processes, *Sustainable (Bio) Chemical Process Technology*, pp. 317–324, 2005.

[10] A. Brook, D. Kendrick, A. Meeraus and R. Raman: GAMS - A User's Guide, Washington: GAMS Development Corp., 1998.

[11] H.-P. Schwefel: Evolution and Optimum Seeking, New York: Wiley Interscience, Sixth-Generation Computer Technology, 1995.

[12] H. G. Beyer and H. P. Schwefel: Evolution Strategies - a comprehensive introduction, *Natural Computing: An International Journal 1(1)*, pp. 3–52, 2002.

Syngas Chemical Looping Gasification Process and Its Novel Retrofit Applications

Fanxing Li, Chemical and Biomolecular Engineering, The Ohio State University, Columbus, OH

Deepak Sridhar, Chemical and Biomolecular Engineering, The Ohio State University, Columbus, OH

Luis Gilberto Velazquez Vargas, The Babcock and Wilcox Company, Barberton, OH

Andrew Tong, Chemical and Biomolecular Engineering, The Ohio State University, Columbus, OH

Liang Zeng, William. G. Lowrie Dept. of Chemical and Biomolecular Engineering, The Ohio State University, Columbus, OH

Fei Wang, Chemical and Biomolecular Engineering, The Ohio State University, Columbus, OH

Zhenchao Sun, Chemical and Biomolecular Engineering, The Ohio State University, Columbus, OH

L. -S. Fan, Chemical and Biomolecular Engineering, The Ohio State University, Columbus, OH

The Syngas Chemical Looping (SCL) gasification process co-produces hydrogen and electricity from gaseous fuels such as syngas and light hydrocarbons. In this process, a reactive oxygen carrying medium is first reduced with gaseous fuels and then regenerated with steam and air in a cyclic manner. Through the assistance of the reaction medium, the SCL process simplifies the overall fuel conversion scheme while reducing pollutants in an indirect but effective manner. Extensive demonstrations of the SCL process have been carried out in a bench scale (2.5 KWt) unit using coal derived syngas, natural gas, and other light hydrocarbons as feedstock. More than 99.5% fuel conversion was achieved. The average hydrogen product purity was higher than 99.95%. A sub-pilot scale demonstration unit has also been constructed and tested. In this paper, the status of the SCL process demonstrations is discussed. Novel retrofit applications of the SCL system for liquid fuel and electricity generations are also explored. Based on ASPEN Plus® analysis, the SCL process can improve the efficiency of existing energy conversion systems by 5 - 10%.

See more of this Session: Advances in Process Intensification I
See more of this Group/Topical: Process Development Division

Two Stage Acid Hydrolysis of Agricultural Waste Using Microwave Reactor Technology

Hariprasad Amanapu, Chemical Engineering, Clarkson university, Potsdam, NY

Roshan Jachuck, Process Intensification & Clean Technology (PICT) Group, Clarkson University Center for Advanced Materials Processing, Potsdam, NY

The alarming rate of fossil fuels depletion has increased the concern for alternative energy sources like bio-ethanol, bio-diesel, fuel cell, solar energy etc. Cellulosic ethanol offers great promise as an alternative fuel with the availability of 73.9 teragrams agricultural waste [1] around the world every year. Cellulosic ethanol is a technically viable but not a cost effective process because of low yield of glucose during the hydrolysis of cellulose. The use of microwaves is making in roads into organic synthesis with the advantages of high reaction rates, energy savings and reduction in process time .The idea of two stage acid hydrolysis in the presence of microwaves is investigated for maximum yield of glucose. First order kinetics is assumed for both main reaction and side reaction, and activation energy constants are estimated.

References:

[1] Seungdo Kim, Bruce E. Dale , Biomass and Bio-energy, 26 (2004) 361-375.

See more of this Session: Advances in Process Intensification I
See more of this Group/Topical: Process Development Division

Green Diesel Additive Synthesis: Elimination of Azeotropic Distillation by Coupling Simulated Moving Bed Reactor with Solvent Recovery Units

Viviana T. Silva, LSRE - Laboratory of Separation and Reaction Engineering - Faculty of Engineering - University of Porto, Porto, Portugal

Rebeca Silva, LSRE - Laboratory of Separation and Reaction Engineering - Faculty of Engineering - University of Porto, Porto, Portugal

Alirio Egidio Rodrigues, Chemical Engineering, LSRE - Laboratory of Separation and Reaction Engineering - Faculty of Engineering - University of Porto, Porto, Portugal

The use of green additives for diesel oxygenation offers the possibility of reducing emissions significantly, mainly Particle Matter (PM), improving even engines performance. Additionally, with the use of oxygenates fuel blends it is possible to introduce a considerable renewable fraction into the petroleum derived fuels, as recommended by the Directive 2003/30/EC. Acetals are recognized as diesel additives. In fact, 1,1-diethoyxethane (1,1-DEE), also known as diethylacetal, has only been reported as a good diesel additive in the 80's by Daimler Benz [1], and just in the most recent studies [2,3] it has been confirmed that 1,1-DEE reduces drastically particles emissions.

The synthesis of 1,1-DEE involves a chemical equilibrium limited reaction, and the reactive mixture exhibits three binary azeotropes (ethanol/water, ethanol/acetal and acetal/water) and one ternary azeotrope (ethanol/water/acetal), making its production and commercialization highly dependent on energy consumption. Those disadvantages led LSRE researchers to consider the Simulated Moving Bed Reactor (SMBR), which combines reaction and adsorption separation into a single unit, achieving complete reactants conversion and complete products separation [4]. As result, and for the first time, it was applied the SMBR technology for an acetalization process [5]. By this way, it is possible to obtain two outlet streams: (i) the extract stream comprising ethanol and water; and (ii) the raffinate stream comprising ethanol and acetal. Ethanol should be separated from each stream and recycled back to SMBR unit. Since distillation is the most widely used separation process in the chemical industry; the recovery of ethanol from each stream by distillation leads to a binary azeotrope (considering 100% purity of each stream in a base without ethanol). Reducing the purity criteria, the separation of each stream will also lead a ternary azeotrope. To break these azeotropes by selecting a proper entrainer is costly. Therefore, alternative strategies are essential for the economic viability of the process.

This work presents a methodology to avoid azeotropic distillation, and therefore avoiding the use of solvents to break the azeotropes and reducing the associated solvent distillation costs. The proposed methodology uses residue curve map (RCM) in order to define the distillation regions for the system ethanol/1,1-DEE/water and to set the minimum purity requirement of each SMBR outlet stream (raffinate and extract) which will be distillated. Thus, the RCM will enable to design the SMBR under reduced purity requirements in order to reduce the desorbent consumption and consequently the distillation cost. Moreover, a new strategy was developed and implemented to optimize the overall process: elimination of the azeotropic distillation by recycling back to the SMBR the azeotropic mixtures ethanol/ water and ethanol/acetal obtained in the distillation of extract and raffinate streams, respectively.

References:

1. K. Boennhoff, F. Obenaus, "1,1-Diethoxyethane as diesel fuel", CA Patent no. CA1141544A1 (1980).

2. F. Frusteri, L. Spadaro, C. Beatrice, C. Guido, "Oxygenated additives production for diesel engine emission improvement", Chem. Eng. J. 134 (2007), 239-245.

3. K. Nord, D. Haupt, "Reducing the Emission of Particles from a Diesel Engine by Adding an Oxygenate to the Fuel", Environ. Sci. Technol. (2005) 39, 6260-6265.

4. Silva, V.M.T., Rodrigues, A.E., "Novel Process for Diethylacetal Synthesis", AIChE J., (2005) 51, 2752-2768,.

5. A. E. Rodrigues, V. M.T.M. Silva, "Industrial process for acetals production in a simulated moving bed reactor", WO Patent 2005/113476A1 (2005), EP1748974 (2007).

See more of this Session: Advances in Process Intensification I
See more of this Group/Topical: Process Development Division

Supported Ionic Liquid Phase (SILP) Catalysts for Ambient Pressure and Low Temperature Water-Gas-Shift Reaction

Sebastian Werner[a], Marco Haumann[a], Normen Szesni[b], Peter Wasserscheid[a]

[a] *Chair of Chemical Reaction Engineering, University Erlangen-Nürnberg, Erlangen, Germany*
[b] *Süd-Chemie AG, Bruckmühl, Germany*

Abstract: Supported Ionic Liquid Phase (SILP) catalysts are new materials consisting of an ionic liquid, a metal catalyst and a porous support. The catalyst is dissolved in the ionic liquid which itself is dispersed as a thin film on the inorganic support. This application combines both the advantages of homogeneous and heterogeneous catalysis and thus bridges the gap between traditional homogeneous and heterogeneous catalysis. Especially continuous, gas-phase reactions are highly suited for this novel and innovative technology. An industrially important example is the water gas shift (WGS) reaction, by which hydrogen can be generated from carbon monoxide and water. Homogeneous WGS catalysts operate at milder temperatures than commercial heterogeneous systems. This can open new fields of applications such as distributed hydrogen production or fuel reforming. Since hydrogen production via WGS is an exothermic reaction, lower temperatures result in higher equilibrium conversions. In this contribution we present investigations of homogeneous metal complexes active in the water gas shift reaction which have been immobilized by the SILP technique. In a first set of experiments known homogeneous systems and newly developed complexes were screened in a continuous rig at temperatures of 100 to 160°C and 1 to 10bar. The most active complexes were investigated in detailed kinetic studies including variation of reaction temperature, partial pressures of CO and H_2O, residence time, variation of the ionic liquid, support and catalyst loading. The investigated SILP systems exhibit activities and stabilities exceeding those of homogenous systems reported in literature. Even commercially available WGS catalysts are outperformed by the newly developed catalysts under the applied temperature conditions. The results indicate that SILP derived WGS catalysts may become a promising alternative to conventional heterogeneous systems, especially in distributed fuel cell setups or decentralized biogas applications.

Keywords: Water Gas Shift Reaction; Ruthenium; Homogeneous Catalysis; Immobilization; Supported Ionic Liquid Phase

1. INTRODUCTION

1.1 Supported Ionic Liquid Phase (SILP) catalysis.

The immobilization of homogeneous catalysts by ionic liquids has been studied extensively during the last decade. The ionic nature, very low volatility and thermal stability of ionic liquids make them highly suitable for biphasic ionic liquid-organic liquid transition metal catalysis (Haumann, 2007). The almost unlimited pool of anion/cation-combinations allows the synthesis of tailor made ionic liquids that are capable of stabilizing the catalytically active species (Wasserscheid, 2000). Moreover, appropriate ionic liquids can enhance substrate solubility while possessing low product solubility, and thus facilitate simple product separation and complete catalyst recovery (Wasserscheid, 2001).

By dispersing the ionic liquid on the large surface of a porous material a heterogenized form of a homogeneous catalyst, a so-called Supported Ionic Liquid Phase (SILP) system, is obtained as depicted in Figure 1 (Riisager, 2005). These SILP systems are well suited for continuous gas phase reactions, since the large surface area can minimize mass transport limitations. In addition, the high dispersion of the ionic liquid catalyst solution on the support surface results in a very efficient use of the ionic liquid and the precious catalyst.

The solid SILP catalysts allow the use of fixed-bed reactor technology and therefore combine both the advantages truly bridging the gap between traditional homogeneous and heterogeneous catalysis.

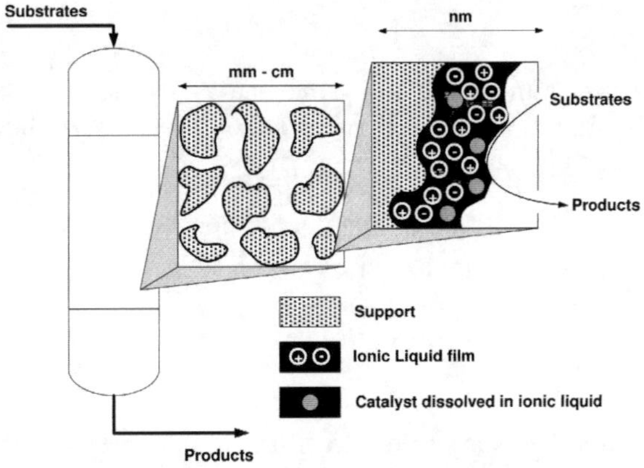

Fig. 1. Schematic representation of Supported Ionic Liquid Phase (SILP) systems for catalysis.

Especially continuous, gas-phase reactions are highly suited for this novel and innovative catalyst technology. The SILP concept has been successfully implemented in continuous gas phase hydroformylation of C3 and C4 feed stocks (Haumann 2007). Classical homogeneous rhodium catalysts exhibited high activity and selectivity accompanied by excellent long term stability with more than 800 hours time on stream (Wasserscheid, 2005).

1.2 Water Gas Shift (WGS) Reaction.

An further industrially relevant example for the successful application of SILP catalysts is the water gas shift (WGS) reaction, by which hydrogen can be generated from carbon monoxide and water. Since hydrogen production via WGS is an exothermic reaction, lower reaction temperatures result in higher equilibrium conversions (Laine and Crawford, 1988).

$$CO + H_2O \rightleftharpoons CO_2 + H_2 \qquad \Delta H_R = -41.2\ kJ\ mol^{-1}$$

Standard WGS catalysts are based on base metal oxides or supported precious metals. The use of metal oxides (e.g. Fe_2O_3/Cr_2O_3) requires very high temperatures and therefore results in low equilibrium conversions as displayed in Figure 2.

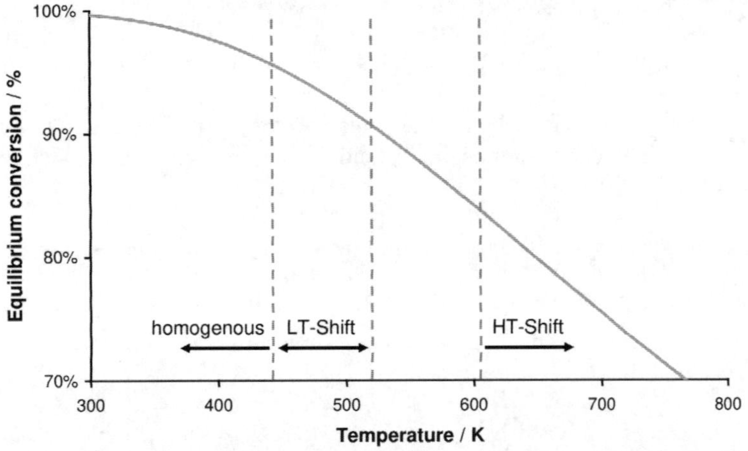

Fig. 2. Calculated equilibrium conversion for the WGSR (AspenPlus2006)

High Temperature Shift (HTS) systems are relatively robust towards poisoning and exhibit good stability. Low Temperature Shift (LTS) catalyst systems (e.g. $CuO/ZnO/Al_2O_3$) operate at milder temperatures, yielding equilibrium conversions of up to 98 % but are more prone to poisoning, in which case extensive gas purification is mandatory (Jacobs and Davis, 2007). The most effective systems so far are combinations of HTS and LTS catalysts. Nevertheless, in order to remove the last traces of CO an additional methanization step is required, affecting negatively process economics.

Homogeneous WGS catalysts are known in the literature since the 1920s and were introduced by Ott (Ott, 1924) and Hieber (Hieber and Becker, 1930). The most active systems are based on ruthenium complexes having activities in the range of 10 $mol_{H2}\ mol_{Ru}^{-1}\ h^{-1}$ at ambient pressure (Ungermann, 1979) and 320 $mol_{H2}\ mol_{Ru}^{-1}\ h^{-1}$ at pressures of 35 bar (Pettit, 1979). Extensive multiphase batch studies indicated that these systems were operating under mass transport limitation even at temperatures around 100 °C (Ford, 1981). All theses studies were conducted in batch or semi-batch setups, thus not investigating stability of those complexes in detail.
Up till now no efficient immobilization strategy has been reported. The multiphase approach resulted in loss of solvent via evaporation whereas direct fixation of the ruthenium complex on a support led to no active catalyst species.

In this contribution we present the efficient immobilization of homogeneous ruthenium WGS complexes in supported ionic liquids on highly porous silica materials. Our approach may open new fields of applications of WGS at very low temperatures, such as distributed hydrogen production or fuel reforming.

2. RESULTS

2.1 Continuous Gas-phase WGS Reaction Setup.

For the evaluation of the homogenous ruthenium complexes the preparation procedure depicted in Figure 3 was followed. First the catalyst $RuCl_3$-hydrate (Alfa Aesar, >99%) was dissolved in water and stirred in a Schlenk-Tube. Afterwards the ionic liquid [BMMIM][OTf] (Solvent Innovation, >99%) was added. After 15 minutes of stirring the highly porous support Silica 100 (Merck) was added. By slow removal of the solvent in vacuum a dry powder with a ruthenium loading of 2 $mg_{Ru}\ g_{Support}^{-1}$ and an ionic liquid loading $V_{IL}\ V_{pore}^{-1} = 0.1$ was obtained.

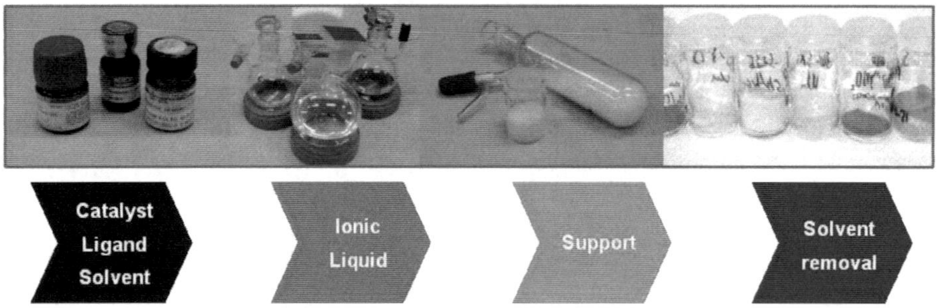

Fig. 3. Preparation procedure of the Supported Ionic Liquid Phase (SILP) catalysts for the WGSR.

The SILP catalysts were filled into a fixed bed tubular reactor rig as displayed in Figure 4. The reaction rig is capable of operation pressures between 1 (atmospheric) and 35 bars by means of a back-pressure valve (Samson) and reaction temperatures of 80 to 250°C.

After heating under constant nitrogen flux (43 ml^N/min) to the desired reaction temperature, dosing of the substrates was started. The volumetric ratio of water (bidestilled) and carbon monoxide (Linde, 3.7) was set to 2:1 so that partial pressures of p_{CO}=0.1 bar and p_{H2O}=0.2bar resulted (balance: nitrogen). A constant residence time in the fixed-bed of τ=2.6s was maintained, which resulted in variable volume flows for nitrogen, carbon monoxide and water depending on the reaction temperature.

The gas-flow leaving the rig was analyzed by an On-Line MicroGC (Varian 4900, MS5A/PPQ) taking samples every 10 minutes.

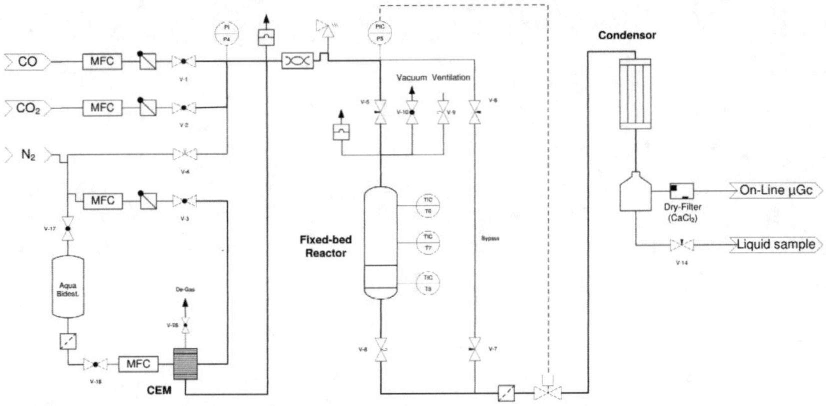

Fig. 4. Flow scheme of the continuous fixed bed reactor rig.

2.2 Parameter screening

In a first set of experiments the reaction parameters such as different reaction temperatures, residence times, partial pressures and total pressure were evaluated. Additionally a variation on catalyst and ionic liquid loading and a variation of the ionic liquid were investigated with respect to the activity of the resulting SILP-catalyst. An overview of different activation profiles obtained in these experiments is depicted in Figure 5.

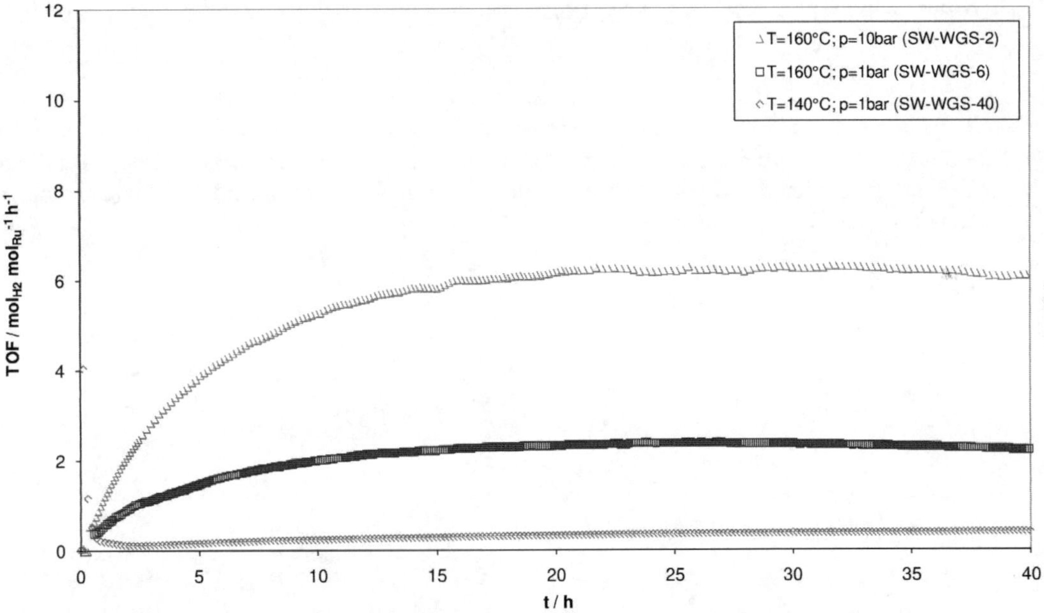

Fig. 5. Parameter screening and resulting Turn-Over-Frequency (mol_{H2} mol_{Ru}^{-1} h^{-1}) of a $RuCl_3$ / [BMMIM][OTf] / Silica-SILP System at p=1..10bar, T=140..160°C, τ=2.6s, p_{CO} = 0.1 bar, p_{H2O} = 0.2bar.

After an initial induction period of 15 to 50 hours depending on Temperature and Pressure a stable operation point is reached. During this time an active species of the catalyst is formed from the $RuCl_3$ precursor.

On long-term studies the system showed serious deactivation tendency that is especially observed at elevated temperatures and high CO pressures. This behavior is depicted in Figure 6 and shows one the major drawbacks of this recently developed system. The loss of activity is directly related to stripping of volatile ruthenium carbonyle compounds.

Besides that, with feed-gases resembling a technical composition (e.g. H_2, CO, CO_2, CH_4) significantly lower activities are observed.

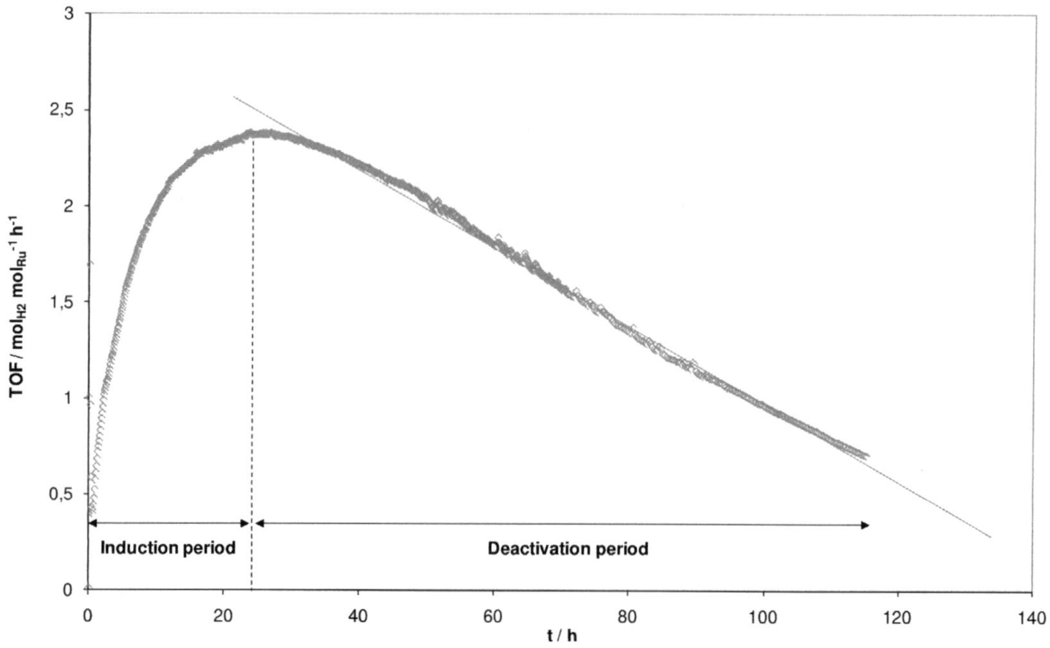

Fig. 6. Activation and Deactivation behavior of a $RuCl_3$-SILP System. Turn-Over-Frequency (mol_{H2} mol_{Ru}^{-1} h^{-1}) of a $RuCl_3$ / [BMMIM][OTf] / Silica-SILP System at p=1bar, T=160°C, τ=2.6s, p_{CO} = 0.5 bar, p_{H2O} = 0.5bar

2.3 Kinetic studies

After the evaluation on activation behavior and induction period, kinetic measurements on activation energy and partial reaction orders in respect to carbon monoxide and water have been carried out under stable operation conditions as depicted in Figure .

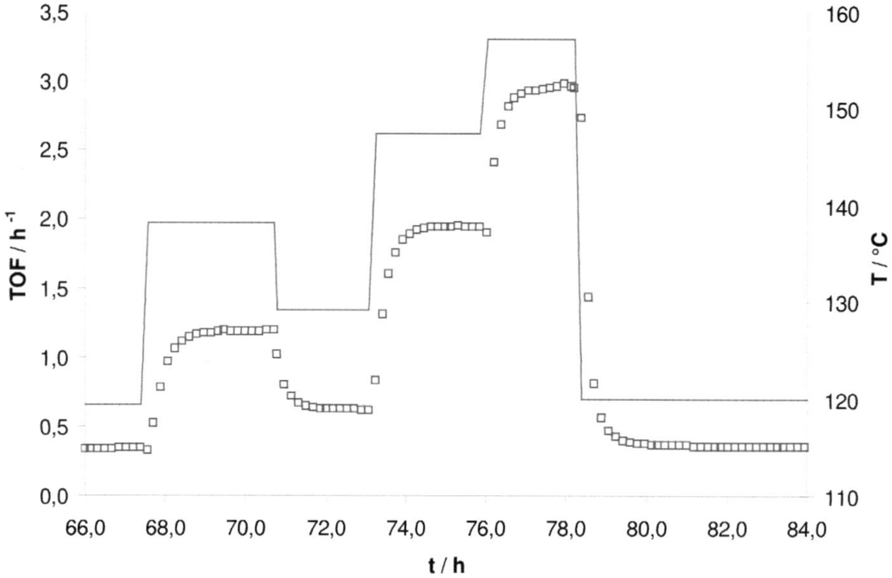

Fig. . Temperature variation and Turn-Over-Frequency (mol_{H2} mol_{Ru}^{-1} h^{-1}) of a $RuCl_3$ / [BMMIM][OTf] / Silica-SILP System at p=1bar, T=120..150°C, τ=2.6s, p_{CO} = 0.1 bar, p_{H2O} = 0.2bar

During a single experimental run a whole set of stable operation points can be measured resulting, and in this way a fast determination of the effective kinetics relevant for detailed reaction engineering is possible. Activation energy and partial reaction orders have been successfully calculated from this data in good agreement with existing literature thus underlining the homogenous nature of this immobilized catalyst.

3. CONCLUSION

We successfully immobilized homogeneous ruthenium based WGS catalysts by the use of Supported Ionic Liquid Phase (SILP) technology. Catalyst activities of more than 6 mol_{H2} mol_{Ru}^{-1} h^{-1} have been realized, which is much higher than any previously published activity for a homogenous WGS system at atmospheric pressure so far. The high activity was maintained over more than 100 hours time on stream, clearly indicating the significant technical potential of these novel catalytic materials for continuous WGS processes.

The SILP concept is most advantageous for continuous gas-phase reactions, where the combination of well-defined catalyst complexes, low-volatile ionic liquids and solid, porous supports can enhance the process economics with respect to equilibrium conversion, mass transport limitations and product/catalyst recovery. We therefore anticipate that the SILP catalysis concept will help bridging the gap between homogeneous and heterogeneous catalysis and will lead to improved catalytic processes in the near future.

Nevertheless, two major drawbacks of the hitherto prepared SILP-catalysts are the deactivation due the stripping of ruthenium carbonyle complexes and the low (to negligible) activity when applied to industrially relevant feed compositions.

REFERENCES

Ford, P. (1981). The water gas shift reaction: homogeneous catalysis by ruthenium and other metal carbonyls. Acc. Chem. Res., 14, 31-37.

Haumann, M., Riisager, A. (2008). Hydroformylation in Room Temperature Ionic Liquids (RTILs): Catalyst andProcess Developments. Chem. Rev., 108, 1474.

Haumann, M., Dentler, K. Joni, J. Wasserscheid, P. (2007) Continuous Gas-Phase Hydroformylation of 1-Butene using Supported Ionic Liquid Phase (SILP) Catalysts. Adv. Synth. Catal. 349, 425.

Hieber, W. and Becker, E. (1930). Metal carbonyls. VIII. Iron tetracarbonyl. Chem. Ber., 63, 1405.

Jacobs, G. and Davis, B.H. (2007). Low temperature water -gas shift catalysts. Catalysis, 20, 122–285

Laine, R.M. and Crawford, E.J. (1988). Homogeneous catalysis of the water-gas shift reaction. J. Mol. Catal., 44, 357.

Ott, S. (1924). Pyrogenic dissociation of sodium oxalate. Zeitschr. Physik. Chem., 109, 1.

Pettit, R., Cann, K., Cole, T., Mauldin, C.H. and Slegeir, W. (1979). Catalytic reductions using carbon monoxide and water in place of hydrogen. Adv. Chem. Series, 173, 121.

Riisager, A. Fehrmann, R., Flicker, S., van Hal, R., Haumann, M., Wasserscheid, P. (2005) Very Stable and Highly Regioselective Supported Ionic-Liquid-Phase (SILP) Catalysis: Continuous-Flow Fixed-Bed Hydroformylation of Propene. Angew. Chem. Int. Ed. 44, 815.

Riisager, A., Fehrmann, R., Haumann, M. Gorle, B.S.K., Wasserscheid, P. (2005), Stability and Kinetic Studies of Supported Ionic Liquid Phase Catalysts for Hydroformylation of Propene. Ind. Eng. Chem. Res. 44, 9853.

Ungermann,C , Landis, V., Moya, S. A., Cohen, H., Walker, H., Pearson, R.G., Rinker, R.G. and Ford, P.C. (1979). Homogeneous catalysis of the water gas shift reaction by ruthenium and other metal carbonyls. Studies in alkaline solutions. J. Am. Chem. Soc., 101, 5922.

Wasserscheid, P., Keim, W. (2000). Ionic Liquids – New "Solutions" for Transition Metal Catalysis. Angew. Chem. Int. Ed., 21, 3772.

Wasserscheid, P., Eichmann, M. (2001). Selective dimerisation of 1-butene in biphasic mode using buffered chloroaluminate ionic liquid solvents—design and application of a continuous loop reactor. Catal. Today, 66, 309.

Solar-Thermal Reactor Engineering - the Ultimate in Process Intensification

Paul Lichty, Department of Chemical and Biological Engineering, University of Colorado, Boulder, CO

Janna Martinek, Department of Chemical and Biological Engineering, University of Colorado, Boulder, CO

Bryan Woodruff, Department of Chemical and Biological Engineering, University of Colorado, Boulder, CO

Elizabeth Saade, Department of Chemical and Biological Engineering, University of Colorado, Boulder, CO

Carl Bingham, National Renewable Energy Laboratory, Golden, CO

Alan W. Weimer, Department of Chemical and Biological Engineering, University of Colorado, Boulder, CO

High-temperature solar-thermal reactor engineering has been a focus at the University of Colorado for over a decade. Three generations of solar reactors have been developed and demonstrated on-sun in the High-flux Solar Furnace (HFSF) at the National Renewable Energy Laboratory (NREL). Reaction temperatures exceeding 1000 deg C are achieved using concentrated sunlight.

Among the processes being developed are the production of syngas from biomass gasification, the dissociation of methane to produce hydrogen and carbon black, and various metal oxide reduction reactions as the high-temperature step in multi-step thermochemical water-splitting cycles.

This paper will review the fundamental thermodynamic considerations for solar-thermal processing and recent results for the rapid conversion of biomass by pyrolysis and gasification. The process provides for a higher yield of syngas from biomass compared to conventional thermochemical gasification, an increase in the heating value of the product sygnas vs. the starting biomass and a significant reduction/elimination of tar formation.

Solar field design will also be presented - i.e. mirror array, tower height, reactor materials and design considerations. A particular emphasis will be placed on identifying critical issues that need overcome to commercialize the process.

See more of this Session: Advances in Process Intensification I
See more of this Group/Topical: Process Development Division

Systematic Analysis, Design and Optimization of Heterogeneously Catalyzed Gas Phase Reaction Processes

A. Peschel[1], H. Freund[1], K. Sundmacher[1,2]

[1] Max Planck Institute for Dynamics of Complex Technical Systems, Sandtorstaße 1, 39106 Magdeburg

[2] Process Systems Engineering, Otto-von-Guericke University Magdeburg, Universitätsplatz 2, 39106 Magdeburg

In order to determine the optimal reaction conditions of a process, most often heuristics, attainable region approaches or rigorous optimization methods are applied. One of the most successful methods for solving the reactor synthesis problem is the rigorous optimization of a reactor superstructure. However, the obtained networks consist usually of several connected ideal standard reactors, such as CSTR and PFTR. Due to the large number of reactors required to build such networks, the obtained solutions can hardly be realized in practice. In addition, not all process intensification measures such as the distributed dosing of a reactant, increasing the area for heat transfer or the interface area in a multiphase reaction systems can be investigated easily with the aforementioned methods.

In this contribution, we propose a new methodology for the analysis, design and optimization of chemical reactors. This approach is not based on the use of standard reactors. It is intended for the investigation of all kind of process intensification measures, for example to obtain an optimal temperature profile in the reactor.

The method is a multi-level approach. On the first level, the investigation is based on tracking a fluid element on its way through the process [1]. The state of the fluid element at the reactor inlet is fixed. At the reactor outlet specific states are constrained and measures for the reactor performance, such as the conversion, are enforced. The task is to find the optimal way of the fluid element with respect to a certain objective, such as minimum residence time or maximum selectivity. In order to optimize the process route, certain process variables, which are taken to be variable over time, are optimized. Hence, the fluid element is optimally manipulated at every time along its path through the reactor. On this level of the proposed methodology, only system inherent limitations, for example constraints which are imposed by the catalyst, are considered. Technical limitations due to the reactor design are not yet included in the model. This gives rise to the maximum potential, which can be obtained by controlling the investigated process variable. In terms of process intensification, the effect of several measures such as perfect heat control using a microreactor or distributing a reactant over the reactor length can be investigated without extensive modeling of all technical constraints.

On the next level of the proposed method, the model is converted to a reactor model with ideal flow field (no backmixing) and technical limitations for heat and mass transfer are considered. By comparing the different optimization levels the loss in the objective by these technical limitations can be quantified.

For every level, a dynamic optimization problem has to be solved to obtain the optimal process route. The balance equations of the problem are discretized to yield a large scale NLP problem. This problem is solved using state-of-the-art NLP solvers, such as *IPOPT* and *CONOPT*. It consists of equality and inequality constraints as well as initial and final conditions. The equality constraints are the discretized balance equations and the kinetic expressions for the reaction rate and for the heat and mass transport. Technical limitations impose most often additional inequality constraints.

In order to illustrate this method, the heterogeneously catalyzed gas phase oxidation of SO_2 to SO_3 is investigated. Sulfur trioxide is one of the major bulk chemicals and is being used to produce sulfuric acid. For the intensification of the SO_2-oxidation process the energy efficiency and the reduction of investment costs are the most important aspects.

Considering this example, several variables affecting the reaction rate can be identified. In this reaction system, the reaction temperature is the most important variable, but other variables such as the partial pressure of SO_2 and O_2 are also important. In order to control the fluid temperature at every point in the reactor, the heat flux must be adjusted locally. This can be achieved by manipulating the local values of the heat transfer area, the heat transfer coefficient or the temperature difference between cooling media and fluid. One or more of these variables must be adjusted along the reactor to obtain the optimal heat flux at every point. For example, the heat exchange area can be changed locally be using a non-conventional reactor design. A non-uniform coolant temperature can be obtained by changing the cooling strategy from a constant cooling temperature, e.g. by using a boiling liquid as coolant, to a counter-current cooling strategy. By dividing the reactor into a few segments and applying different coolant strategies in each segment almost every temperature profile can be achieved in the reactor. In such a way, the optimal temperature profile, which is calculated by solving the dynamic optimization problems, can be approximated. Of course, this approach is not limited to optimizing the temperature profile, but can also be used for other process intensification measures. It yields the optimal process route with respect to the investigated objective and control variables as a benchmark for any technical approximation. In addition, the obtained profiles indicate how the optimal reactor must be designed.

To sum up, the described framework for the analysis, design and optimization of chemical reaction processes is very well suited to determine the quantitative effect of different process intensification measures. The multi-level approach yields the optimal reaction route as well as the influence of technical limitations. On the basis of the optimal route, new reactor concepts for smaller and more intelligent apparatuses can be derived.

Literature:

[1] Freund, H.; Sundmacher, K.; *Chem. Eng. Process.*, 2008, 47(12), 2051.

Countercurrent Cascaded Reactors: Their role in the design of reactive distillation columns

J. K. Cheng, J. D. Ward, and C. C. Yu

Department of Chemical Engineering, National Taiwan University,

Taipei 106, Taiwan

Multifunctional process units are an integral part of process intensification and reactive distillation (RD) is one of the most common examples of a multifunctional processes. Despite recent progress in the feasibility analysis, design, and, in some cases, control of reactive distillation, shortcut design of reactive distillation is not well understood, especially for the design of the reaction zone inside a column. Several authors proposed methods for the feasibility analysis and also for the calculation of minimum–reflux ratio.[1-3] Lee et al. (2003) describes a method to estimate the minimum catalyst loading based on the nonreactive rectification body method, the critical composition profiles, the analysis of eigenvectors of pinch points, and geometrical design insights. Other than this, however, the quantitative description of catalyst loading and distribution is rarely discussed.

In this work, we use the concept of countercurrent cascaded reactors (CCRs) to describe the reactive zone. An analytical expression for the minimum theoretical catalyst loading can be derived as the number of CCRs approaches infinity. Next, McCabe-Thiele type column stepping can be employed to complete the reactive distillation system description.[4] The tradeoff between the catalyst amount and the energy consumption can be found using the proposed methodology and subsequently, a shortcut design procedure can be derived. This shortcut method provides a minimum reflux ratio and the corresponding theoretical catalyst amount so that the number of separation stages and the number of reaction stages can be obtained.

An ideal binary system with the elementary reaction A↔B and ideal vapor liquid equilibrium is used to illustrate the design. Different parameters of the system, including relative volatility, specific reaction constant, equilibrium constant and product purity were evaluated and optimized designs with different parameters were compared to the shortcut design. The sensitivities of different parameters were calculated, and the specific reaction rate was found to have a significant effect on the catalyst amount which affects the calculation of the number of stages in the column. At moderate specific reaction rate, the design calculated by the shortcut method has a good agreement with the optimal design. Although this shortcut method does not always work well, it can be used for a preliminary process design or as the initial case for a rigorous optimization.

References:

1. Barbosa, D.; Doherty, M. F., Design and Minimum-Reflux Calculations for Single-Feed Multicomponent Reactive Distillation-Columns. *Chemical Engineering Science* **1988,** *43* (7), 1523-1537.

2. Hoffmaster, W. R.; Hauan, S., Difference points in reactive and extractive cascades. III - Properties of column section profiles with arbitrary reaction distribution. *Chemical Engineering Science* **2004,** *59* (17), 3671-3693.

3. Lee, J. W.; Bruggemann, S.; Marquardt, W., Shortcut method for kinetically controlled reactive distillation systems. *Aiche Journal* **2003,** *49* (6), 1471-1487.

4. Lee, J. W.; Hauan, S.; Westerberg, A. W., Graphical methods for reaction distribution in a reactive distillation column. *Aiche Journal* **2000,** *46* (6), 1218-1233.

Mass Transfer Characterisitics of a Novel Gas-Liquid Contactor

Gangadhara Raja Muthinti, Chemical Engineering, Clarkson University, Potsdam, NY

Roshan J.J Jachuck, Process Intensification & Clean Technology (PICT) Group, Department of Chemical Engineering, Clarkson University, Potsdam, NY

Rotating horizontal cylinder(RHC) is one of the upcoming intensified modules in the field of process intensification. There is not much literature available on the mass transfer studies in a rotating tube. Thin hydraulic path lengths are obtained in a rotating tube in comparison to the conventional mass transfer equipments. This enhances the rate of mass transfer by orders of magnitude which is one of the important aspects of process intensification. The main aim of this paper is to determine the mass-transfer characteristics of the horizontal rotating tube in continuous mode and compare it with other intensified modules and traditional gas-liquid contactors. Experiments have been carried out with deionized water as liquid phase and pure nitrogen as the gas phase to study the stripping of dissolved oxygen from liquid to gas phase.

See more of this Session: Advances in Process Intensification II
See more of this Group/Topical: Process Development Division

HELIX REACTOR: GREAT POTENTIAL FOR FLOW CHEMISTRY

P. Geerdink, A. van de Runstraat, E.L.V. Goetheer, C.P.M. Roelands

TNO Organisation for applied scientific research
Department of separation technology
Schoemakerstraat 97
P.O. Box 6012
2600 JA Delft
Netherlands

ABSTRACT

The Helix reactor is highly suited for precise reaction control based on good hydrodynamics. The hydrodynamics are controlled by the Dean vortices, which create excellent heat transfer properties, approach plug flow and avoid turbulence. The flexibility of this reactor has been demonstrated using a very exothermic, temperature sensitive alkylation reaction, a mechanically delicate emulsion polymerisation reaction and a precipitation reaction. It was demonstrated that the highly exothermic alkylation reaction could be controlled such that no side reactions occured. In the case of the emulsion polymerisation reaction the gentle mixing conditions of the Helix reactor were used to control the particle size and size distribution. Also for the precipitation reaction it was possible to obtain a narrow particle size distribution and to control the particle size. It can be concluded that the helix reactor is a versatile tool to perform reactions that are typically batch wise operated in a continuous way, and thereby reducing costs and improving safety and quality.

Keywords: Helix reactor, Dean vortices, flow chemistry, process intensification, alkylation, emulsion polymerisation, precipitation, exothermic reaction, hazardous reaction

1. PROCESS INTENSIFICATION

A significant number of industrial processes are based on process designs that date back decades. Often generating significant amount of waste material and consuming large amounts of energy. Therefore, there is a need for improving these processes. The general concept of process intensification can be an interesting route for the processes of the 21st century. One of the main driving forces is the desire to transform inefficient batch processes towards continuous processes, leading to molecular and energy efficient processes. Especially, potentially dangerous reactions can benefit from this paradigm shift leading to intrinsically safe processes. Microreactors are one of the tools which are attracting considerable attention. One of the reasons for this is the high surface area versus volume ratio which enables a good temperature control. Another reason is the relatively small volume of the reactor which, in case of leakage, leads to the loss of only very small amounts of potentially harmful components. However, the throughput per microreactor is relatively low, the scaling up by numbering up is sometimes not economically attractive, and the presence of solids can lead to malfunction of the equipment. In this work the Helix reactor is presented as a reactor type, where efficient energy and mass transfer can occur. It is demonstrated that this type of reactor does not have the drawbacks encountered in microreactors.

2. INTRODUCTION TO THE HELIX REACTOR

A number of years ago TNO developed a heat exchanger with the shape of a twisted tube (Walpot, 2003), which has excellent heat transfer characteristics. This reactor was named the Helix heat exchanger. The Helix reactor is a continuous tubular reactor with very good heat transfer characteristics, as compared to straight tubular reactors (see Figure 1). This device was further improved and adjusted to make it possible to perform continuous reactions (Naphon and Wongwises, 2006). Other advantages of the Helix reactor are the good mixing characteristic and the good plug flow nature (Nauman, 1976). In Figure 1 four situations are compared. It can be concluded that it is possible to operate the Helix reactor at a high Reynolds number and at a high Nusselt number. The main consequence of this behavior is that turbulent flow is avoided, which results in a low pressure drop. This is a major drawback of straight tubes. The main drawback of mini or microreactors is that this type of reactor suffers from a low throughput.

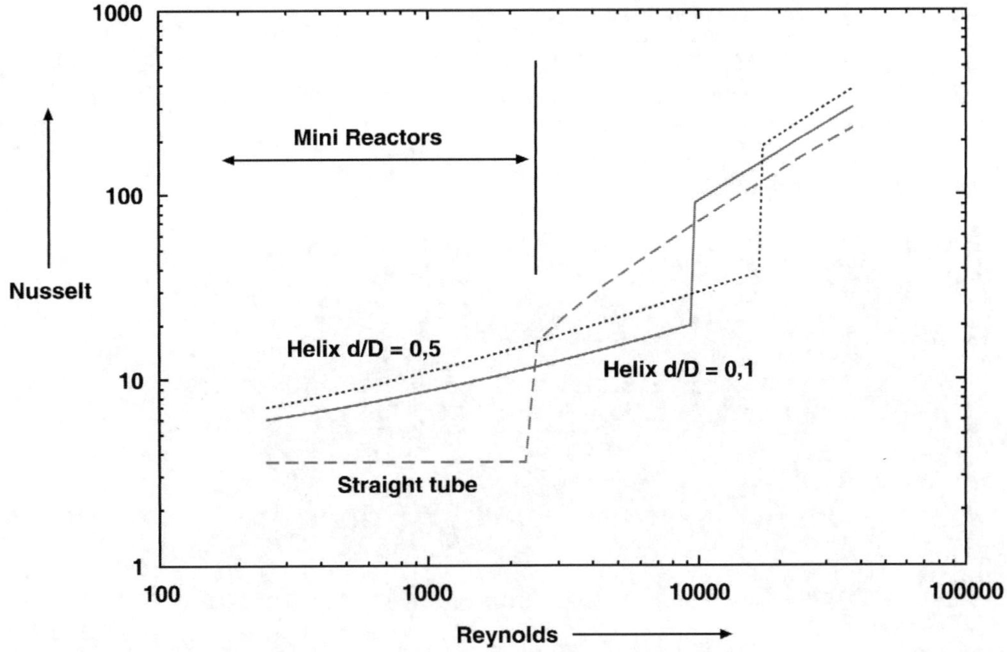

Figure 1. The Nusselt number as function of the Reynolds number. A comparison is made between the Helix reactor and a normal tubular reactor.

The increased heat transfer is caused by Dean vortices as illustrated in Figure 2. These vortices are secondary flows in radial direction at laminar flow speeds. These vortices stabilise the flow so that turbulent properties are formed at higher Reynolds (Hüttl and Friedich, 2000). This results in a flow characteristic that approaches plug flow and a low pressure drop in the tube. It is clear that because of the increased mixing properties of the Helix reactor, this equipment can be suited for continuous reactions.

Figure 2. Dean vortices moving from the wall to the centre of the tube and vise versa.

Process intensification using the Helix reactions.

As a result of the plug flow nature of the Helix reactor, there is narrow distribution of the retention time and this results in a high product quality. The Helix reactor is especially suited for highly exothermic reactions and reactions concerning harmful components. In this work the potential of the Helix reactor is exemplified using three different cases. Namely, a highly exothermic alkylation reaction, an emulsion polymerisation and a reactive crystallisation.

3 CASE 1: SOLVENT FREE CONTINUOUS ALKYLATION REACTION

The synthesis of 1-ethyl-3-methyl-imidazolium bromide ([EMIM] [Br]) is from an industrial perspective an interesting model reaction. [EMIM] [Br] is an ionic liquid that is produced from 1-methylimidazole (MIM) and ethylbromide (EtBr). The reaction between MIM and EtBr is a highly exothermic and hazardous reaction, with an estimated adiabatic temperature rise of 345 °C. Traditional method of production involves dissolving both reactants 20 times, perform the reaction in a stirred tank, and to remove the solvent by distillation after the reaction is complete.

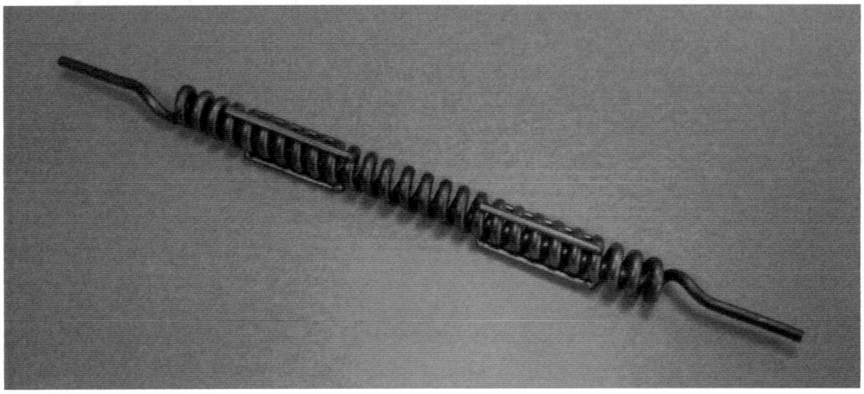

Figure 3. Helix reactor: for this example the distance between to helices, called the pitch, is equal to 12 mm. It is also possible to place this module inside a shell, to obtain a tube-in-shell configuration.

For the Helix setup, two separate pressurised vessels were used one vessel contained EtBr and the other contained MIM. The vessles were pressurised to approximately 6 bars, to assist the pumps to achieve the desired pressure and flow conditions to enter the Helix reactor. The helices had an inner diameter of 2.4 mm, a pitch of 12 mm, and the reactor was made of stainless steel (Figure 3). The Helix reactor was inserted in a shell to optimise the contact with the heat exchanging fluid. Temperature of the reaction medium was measured at 70% of the total length of the helix and before the collection vessel. The collection vessel was kept at 5.5 bars to keep EtBr from boiling. The tubing from the final helix (of the reactor) to the collection vessel was insulated and heated using tracing to keep the product from crystallising inside the tube. Gear pumps and corriolis flow metres were used to control the flow of EtBr and MIM.

It was found that optimal process conditions were 87°C and 5 bar pressure in a Helix reactor with a total length of 21 m. The optimal flow rate was 7.8 kg/h, resulting in a retention time of approximately 113 seconds. For a temperature above 120°C side reactions start to take place, as a result the color of the product becomes yellow to brown, which makes it impossible to sell the material.
In Figure 4 the final product is displayed showing that the reaction is completed. The 11 samples in Figure 4 show that the temperature was controlled to be in the desired temperature range of 87 to 120 °C, and that as a result a white crystalline product was obtained.

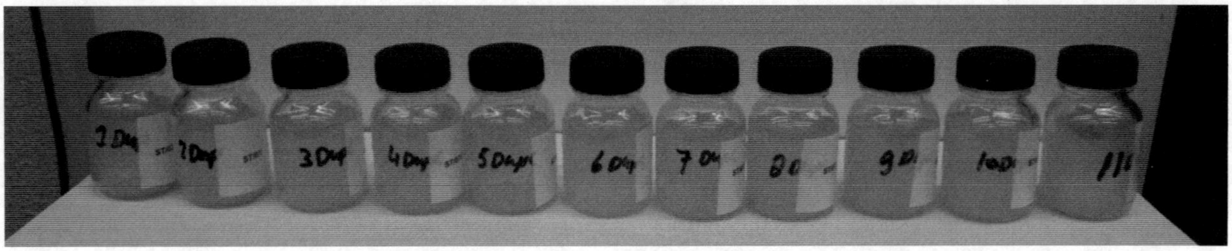

Figure 4. White crystalline [EMIM][Br] produced with the Helix reactor. The product was obtained during 20 minutes of reaction.

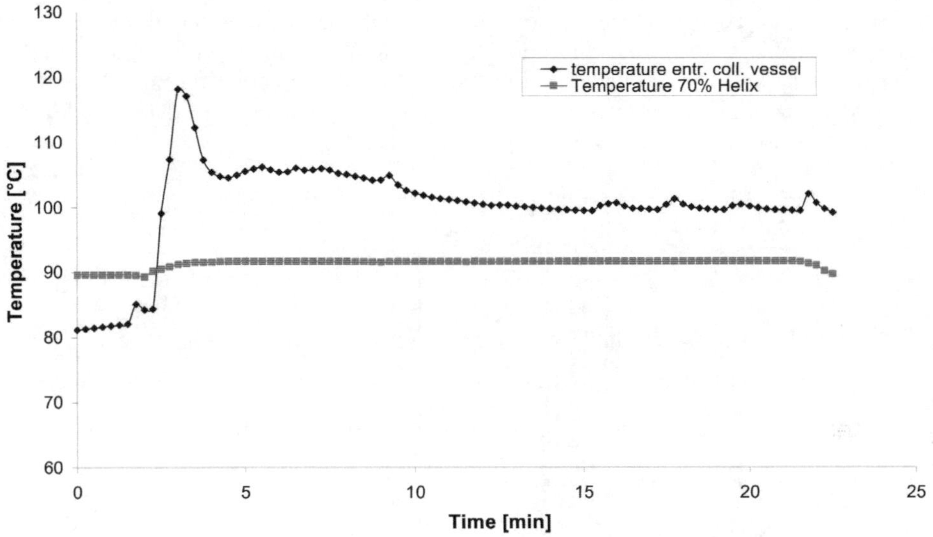

Figure 5. Results of a typical experiment with the Helix reactor to produce [EMIM][Br]. Temperature was measured at 70% of the Helix reactor and at the entrance of the collection vessel.

The temperature at the entrance of the collection vessel is of key importance in using the Helix reactor. As a result of the absence of cooling capabilities in the tubing from the Helix reactor to the collection vessel, an incomplete reaction will result in thermal runaway in this part of the reactor and this will create a yellow to brown colour. During the reaction the temperature stays well below 120 °C indicating the reaction is completed within the Helix reactor, see Figure 5. In fact, besides the start up, the temperature is more or less constant during the full length of the experiment. For both locations the temperature lies between 90°C and 105°C. The obtained product started to crystallise minutes after a sample was taken at high temperature. This indicated a high purity of the product. After crystallisation, there was a fraction in each vessel that did not crystallise. The fraction of reactants that was not converted made out less then 10% of the total volume of the sample. HPLC analyses showed a conversion for all samples of 90% or more, when a residence time of 113 seconds was applied.

The temperature inside the reactor could be fully controlled and colorless liquid product could be obtained which solidified into a white product upon standing. Using this cheap and simple piece of equipment a production of up to 100 tonnes per year can be reached.

4. CASE 2:EMULSION POLYMERISATION USING THE ALTERNATING HELIX REACTOR

A typical emulsion polymerisation process consists of two steps. The first step is the formation of a monomer emulsion and the second step is the reaction from monomer droplet to polymer particle retaining the size of the dispersed phase. The challenges in the second step are avoidance of particle growth and the avoidance of blockage due to polymer deposits. The transition from monomer to polymer goes via a very sticky intermediate which causes adhesion to the wall and particle growth when collisions are sufficiently intense. Prevention of particle growth is performed in the Helix reactor by reducing the difference in speed between particles. Dean vortices result in a plug flow that moves particles both at the wall as in the middle of the tube at the same speed. The absence of internals prevent adhesion of particles inside the reactor, and Dean vortices keep the fluid at the wall moving and in this way adhesion of particles to the wall is prevented.

As an example, poly methylmethacrylate (p-MMA) was produced from 35% methylmethacrylate (MMA), 4 % hexadecane, and using 5 % emulsifier (SDS) and 1 % initiator azo iso-butyronitril (AIBN). The concentration of the emulsifier and the initiator were relative to the concentration of MMA monomer. After an emulsion of the monomer was created using ultrasound, the initiator was added and the mixture was introduced into the Helix reactor. The helices had an inner diameter of 2.4 mm and a pitch of 12 mm, and were made of glass. The helix reactor was designed as a U-shape with a total length of 15 meters, of which only the bottom was filled with liquid. Subsequently the fluid was alternatingly moved back and forth through the reactor using an overpressure. The fluid moved in one direction during 20 seconds before the flow direction was reversed. The optimal liquid velocity in this system proved to be 25 cm/s. After 20 minutes at 75°C conversion was completed and the reaction was quenched.

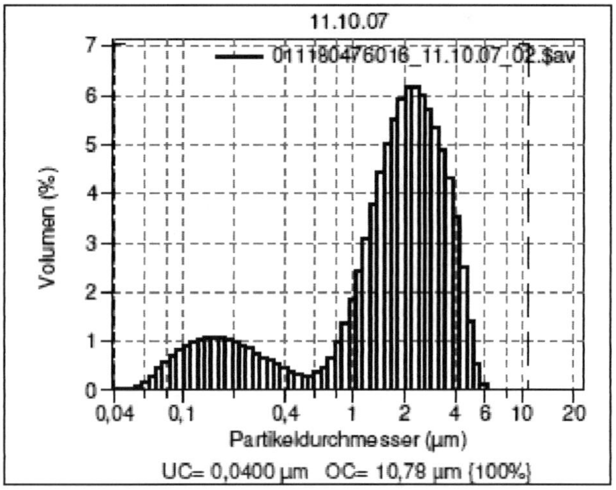

Figure 6. Particle size distribution after polymerisation in a straight tube.

Figure 6 displays the particle size of the polymerised particles produced in a straight tube reactor using the same flow conditions, as used in the Helix reactor. The graph shows that in a straight tube a large particle size distribution is obtained and that on average the particle size is >2µm. Microscopic observation of the product produced in a straight tube reactor indicated that besides agglomerates, the individual particles were also large. A possible explanation is that collision takes place during the entire reaction. Particles that collide in an early stage of the polymerisation result in larger particles, because the monomer is a liquid. Collosions at the end of the polymerisation reaction results in agglomerates since the emulsion droplets are becoming more viscous.

In Figure 7 the particle size distribution is displayed (vol. %) after polymerisation of the emulsion was performed in the alternating Helix reactor. The conversion was complete, there was less than 1% monomer left in the product, and the particle size distribution was greatly improved as compared to the same reaction, performed in a straight tube reactor (Figure 6). The average particle size obtained in the Helix reactor was approximately 70 nm.

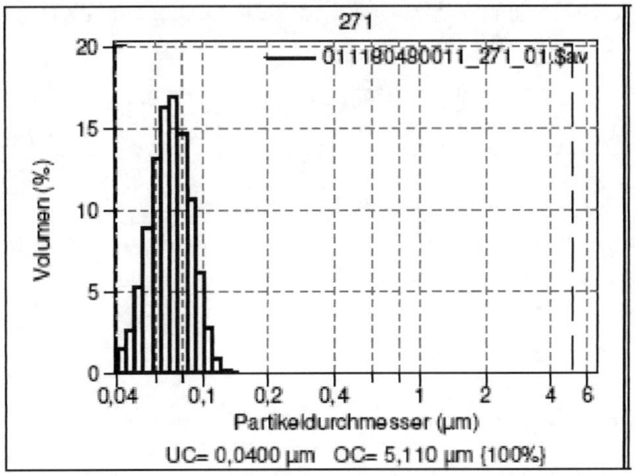

Figure 7. Particle size distribution after polymerisation in the alternating Helix reactor.

5. CASE 3: PRECIPITATION

In the field of particle engineering good control of particle size and particle size distribution is a sine qua non. Continuous stirred reactors can lead to a wide distribution of particle size that may negatively affect product quality. Reactors which resemble ideal plug flow conditions can be a beneficial tool for obtaining monodisperse and/or sub-micron particles. The Helix reactor has been tested for its ability to control size and morphology of precipitated calcium carbonate ($CaCO_3$). Figure 8 displays how flow conditions can influence crystal size and shape.

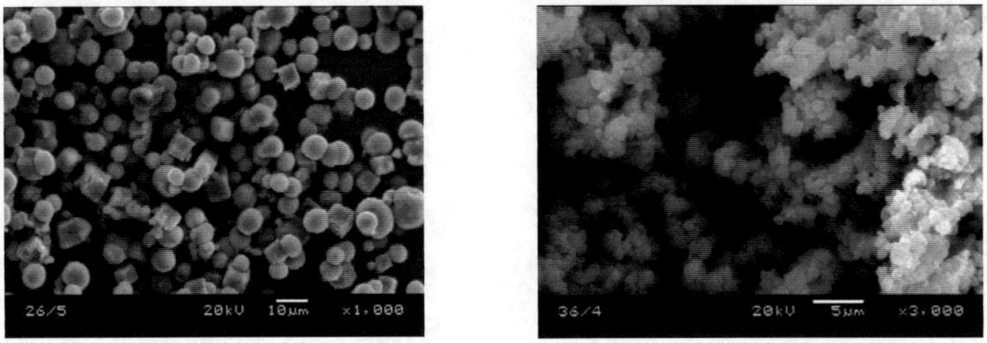

Figure 8. Influence of flow conditions on crystal size and shape; reaction in a straight tube (left), and reaction in the Helix reactor (righ).

Before entering either the Helix reactor or the straight tube reactor a solution of $CaCl_2$ was mixed with Na_2CO_3 in a premixer (Y or T-shaped) followed by further reaction in the reactor. Crystals were obtained and analyzed by FBRM and SEM. A large variety of experimental conditions have been tested. The general conclusions regarding the particle size and particle size distribution are summarized in Table 1. Based on the the good mixing behavior of the Helix reactor in combination with a minimum back mixing, it can be concluded that this type of reactor is suited for liquid/solid reactions.

Table 1. Influence of flow conditions on particle size and particle size distribution using a Helix reactor and a straight tube.

Parameter		Helix versus straight tube	
Flow	Concentration	Particle size	Particle size distribution
Low	Low	smaller	much narrower
	High	smaller	0
High	Low	0	0
	High	much smaller	smaller

6. CONCLUSIONS

The Helix reactor proves to be a multi-purpose tool that can be designed to suit various continuous processes. It is able to transport vast amounts of heat making it ideally suitesd for highly exothermic reactions while the plug flow characteristic ensures continuous high product quality and low byproduct formation. The Helix reactor possesses a gentle mixing characteristic that allows delicate emulsion polymerization reactions to be performed while preventing particle growth by collision or particle adhesion at the wall. The absence of back mixing in the Helix reactor, makes it a suitable tool for liquid/solid reactions, such as precipitation, which requires a small particle size and a narrow particle size distribution. By changing the size of the channel of the Helix reactor, the production rate can be adjusted to fit a desired amount.

REFERENCES

Hüttl, T.J. and Friedrich, R. (2000) Influence of curvature and torsion on turbulent flow in helically coiled pipes. Int Journal of Heat and fluid flow. 21, 345-353.

Naphon, P. and Wongwises, S. (2004) A review of flow and heat transfer characteristics in curved tubes. Renewable & sustainable energy reviews. 10, 463-490.

Nauman, E.B. (1976) The residence time distribution for laminar flow in helically coiled tubes. Chemical engineering science. 32, 287-293.

Walpot, J. (2003) TNO's work on intensification: practical examples. Journal of chemical technology and biotechnology. 78, 236-240.

USING A 5/8" DIAMETER KARR® COLUMN FOR GENERATING SCALE-UP DATA

Authors:

Donald J. Glatz
Wendy L. Parker

Koch Modular Process Systems, LLC.

Prepared for Presentation at

AIChE National Meeting

Nashville, TN
November 2009

INTRODUCTION

The most accurate method used to design liquid-liquid extraction (LLE) equipment has been via pilot plant testing and empirical scale-up. For the KARR® Column most scale-up data is generated in a 1" diameter pilot column. Production KARR® Columns up to 66" diameter have been designed from pilot plant tests in a 1" diameter column. Typically, 50-100 gallons of feed and solvent are required.

A smaller diameter column would have the obvious benefit of requiring less material for testing. For many years, a 5/8" diameter KARR® Column has been used for initial screening studies. However, production columns have not been designed from the 5/8" diameter column since accurate scale-up correlations have not been developed.

A test program is underway to develop the performance comparison between the 5/8" and 1" diameter KARR® Columns. Previous papers presented at AIChE Meetings covered testing with the Toluene-Acetone-Water system[1] and the Butyl Acetate-Acetone-Water system[2]. Both of these systems are high interfacial tension and relatively high density difference. This paper presents data generated using the Ethyl Acetate-Acetic Acid-Water system, which has lower interfacial tension and density difference. A comparison is made between the 1" and 5/8" diameter columns with regards to capacity, efficiency, and agitation speed.

LIQUID-LIQUID EQUILIBRIUM DATA

Liquid-liquid equilibrium data was generated for the Ethyl Acetate-Acetic Acid-Water system using a standard "Shake Test" procedure[3]. The results are reported graphically in Figure 1. As can be seen, the equilibrium curve is not a straight line and thus the theoretical stages for this system are calculated via the standard McCabe-Thiele method commonly associated with distillation[3]. For each run in the pilot plant column, the acetic acid concentration in the feed, solvent, raffinate, and extract are measured analytically. After plotting the equilibrium curve and the operating line, the stages are stepped off from the feed concentration (X_f) down to the raffinate concentration (X_r).

PILOT PLANT EQUIPMENT SET-UP

A simplified schematic of the extraction pilot plant may be seen in Figure 2. This system contains an extraction column and equipment for feeding aqueous phase (water + acetic acid) and ethyl acetate phase. Also, a variable speed pump was used for the bottoms take-off and interface control.

Both columns consisted of 4' (1.22 m) plate stack height, with 2" diameter expanded ends for disengagement. The 5/8" diameter column had 1" plate spacing and the 1" diameter column had 2" plate spacing. 316LSS plates were used in both columns.

TEST PROGRAM OPERATING PROCEDURE

For all tests in this study, the aqueous phase was continuous and the organic phase was dispersed. Only continuous to dispersed (c-d) direction of mass transfer was considered. The operating procedure was as follows:

1. The column was initially filled with distilled water

2. The flows of aqueous feed containing 5 – 6% acetic acid and fresh ethyl acetate were started and the rates set.

3. The aqueous underflow pump was started and adjusted as required to maintain a constant interface level in the top of the column.

4. The agitation speed was set.

5. For each run, feed and solvent rates were checked and adjusted every 15 - 30 minutes. At the completion of each run, raffinate and extract flow rates were measured and samples of feed, solvent, raffinate and extract were taken.

6. During start-up each day, a total of 5 column turnovers were performed before the initial samples were taken. A column turnover is defined as the total column volume divided by the combined feed and solvent rates. After the first run, and following adjustments, a total of 3 turnovers were performed before sampling each subsequent run.

KARR ®is a registered trademark of Koch-Glitsch, L.P.

RESULTS AND DISCUSSION

A summary of the results from all pilot plant runs has been provided on the attached plots. Data for each column was generated at specific throughputs ranging from 10 to 40 m^3/m^2 hr.

- Figures 3 (A, B, and C) and Figures 4 (A, B, and C) provide the raw data for the tests runs. These curves plot the % acetic acid in the raffinate vs the agitation speed (SPM or strokes per minute) at 10, 20, and 30 m^3/m^2 hr for each column. The 5/8" diameter column could not operate at 40 m^3/m^2 hr due to flooding even without agitation. Only one run was successfully completed at 40 m^3/m^2 hr in the 1" diameter column. This runs was performed at a low agitation speed of 100 SPM. Attempts to operate at higher speeds also resulted in flooding.

- Figure 5A and 5B show the stepwise theoretical stage calculation for the 1" diameter column at 30 m^3/m^2 hr and the 5/8" diameter column at 30 m^3/m^2 hr, respectively. This procedure was repeated for the optimal point (lowest acetic acid concentration in the raffinate) for all Figure 3 and Figure 4 curves, thus generating the data for the Stichlmair[4] Type Plot.

- Figure 6 is the Stichlmair Type Plot for the Ethyl Acetate–Acetic Acid–Water data generated during this test program. It shows Efficiency (stages per meter) vs Capacity (m^3/m^2hr) for both the 5/8" and 1" diameter columns. The efficiency is defined as the number of theoretical stages (calculated by graphical solution) divided by the active height of the column. The data reveals relatively equal performance by both columns.

- As mentioned in the Introduction section, data was previously generated and presented for the systems of Butyl Acetate–Acetone–Water and Toluene–Acetone–Water. The Stichlmair Type Plots for these systems are presented as Figure 7 and Figure 8, respectively. These plots also reveal similar performance by both columns for each of these systems.

CONCLUSIONS

Based upon the data presented in this report, the following conclusions are drawn about the performance of the pilot KARR® Columns:

1. The performance of the 5/8" diameter column shows a strong correlation with the 1" diameter column for the ethyl acetate–acetic acid–water system.

2. Since similar results were generated for two other test systems, the capacity and efficiency for the 5/8" diameter and 1" diameter KARR® Columns can be considered to be the same for clean systems.

KARR ®is a registered trademark of Koch-Glitsch, L.P.

RECOMMENDATIONS

The data from three "clean" systems has shown that the performance with regards to capacity and efficiency of the 5/8" diameter KARR® Column is the same as the performance of the 1" diameter KARR® Column. The next step will be to confirm this result for actual feed and solvent materials generated in either pilot plant or production processes.

LITERATURE CITED

1. Cusack, R., Glatz, D., and Holmes, T., "Performance Characteristics in a 5/8" Diameter Karr Column", AIChE Fall National Meeting, Nov. 12-16, 2000.

2. Glatz, D., and Parker, W., "Performance Characteristics in a 5/8" Diameter Karr Column", AIChE Fall National Meeting, Nov. 16-21, 2003.

3. Cusack, R., Fremeaux, P., and Glatz, D., "A Fresh Look at Liquid-Liquid Extraction, Extraction Systems", Chemical Engineering, February 1991.

4. Stichlmair, J., "Chemie-Ingenieur-Technik", 52, 3 (1980)

KARR ®is a registered trademark of Koch-Glitsch, L.P.

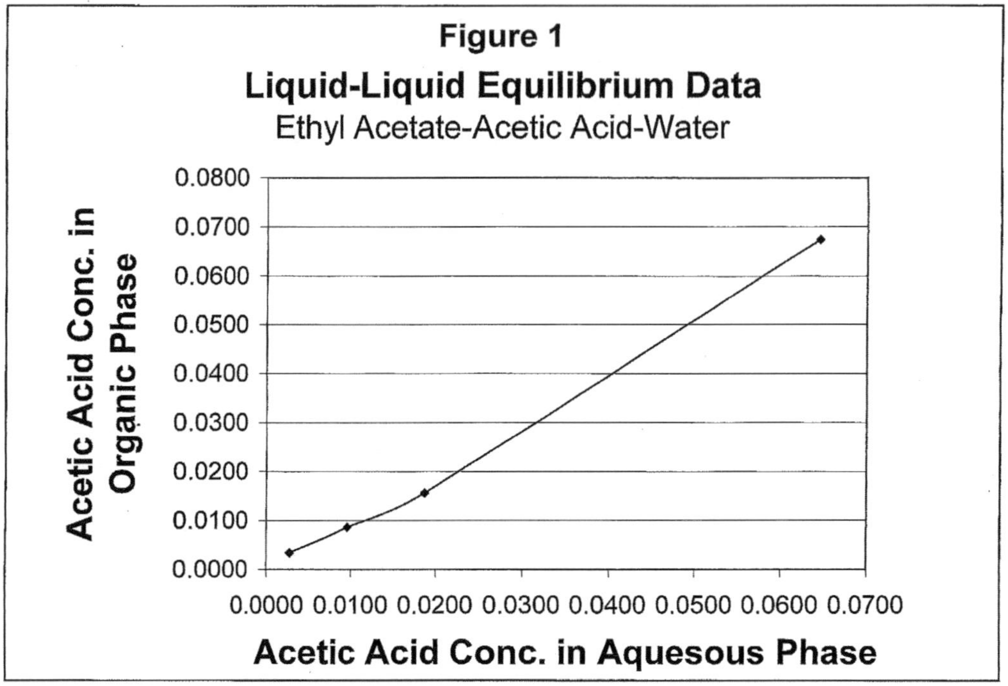

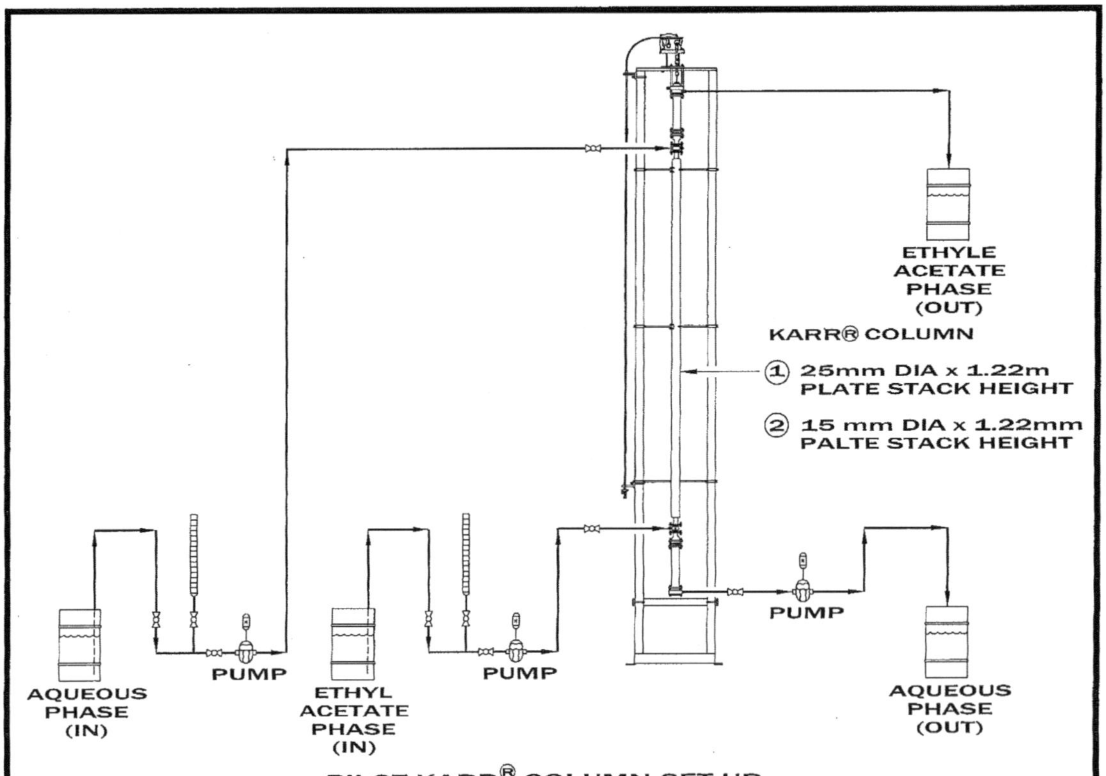

PILOT KARR® COLUMN SET-UP

FIGURE - 2

KOCH MODULAR PROCESS

Figure 3A - Acetic Acid Conc. vs Agitation Speed
1" Diameter Column at 10 m^3/m^2 hr

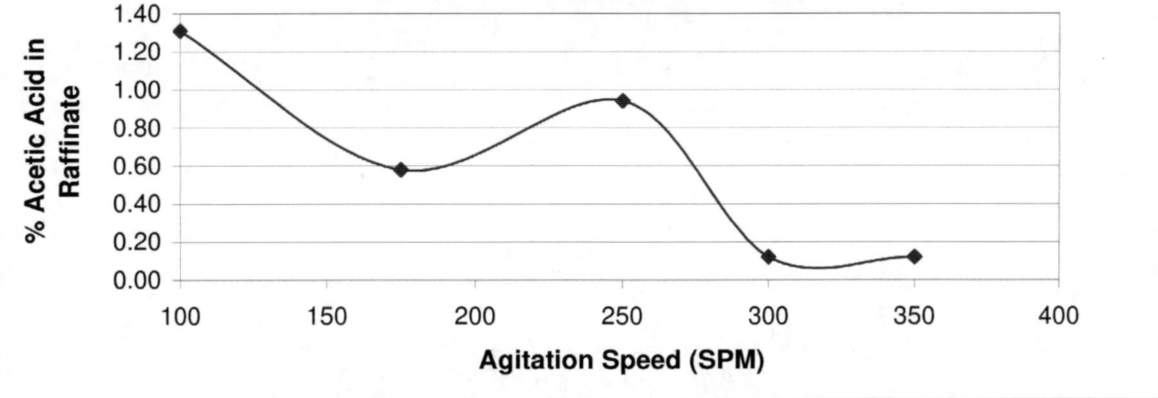

Figure 3B - Acetic Acid Conc. vs Agitation Speed
1" Diameter Column at 20 m^3/m^2 hr

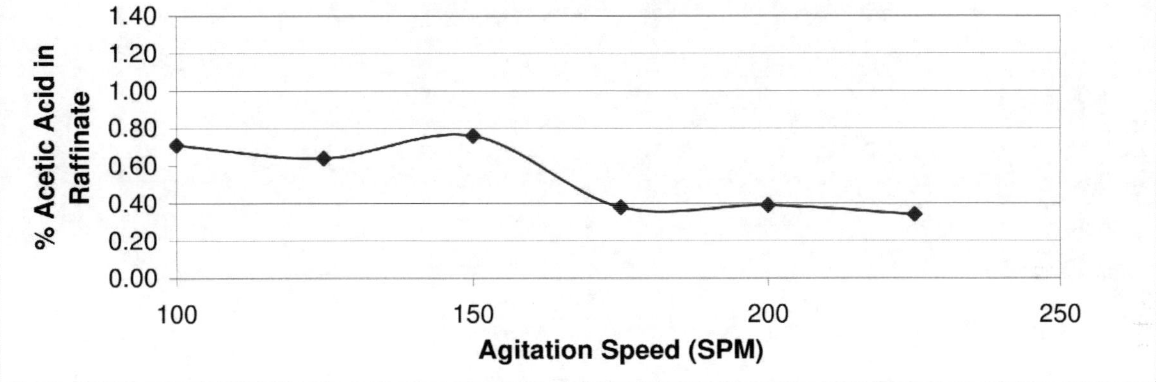

Figure 3C - Acetic Acid Conc. vs Agitation Speed
1" Diameter Column at 30 m^3/m^2 hr

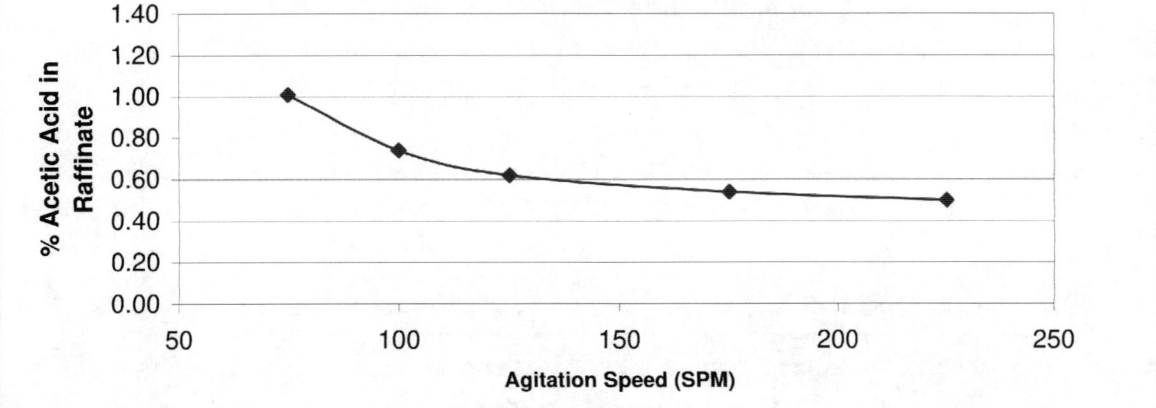

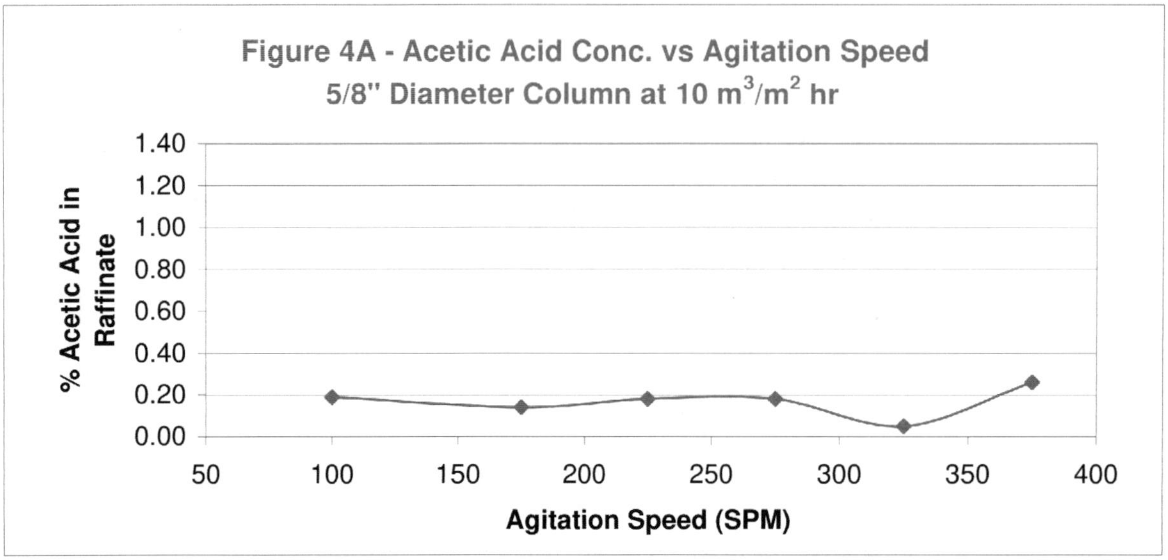
Figure 4A - Acetic Acid Conc. vs Agitation Speed
5/8" Diameter Column at 10 m3/m2 hr
% Acetic Acid in Raffinate
1.40
1.20
1.00
0.80
0.60
0.40
0.20
0.00
50
100
150
200
250
300
350
400
Agitation Speed (SPM)

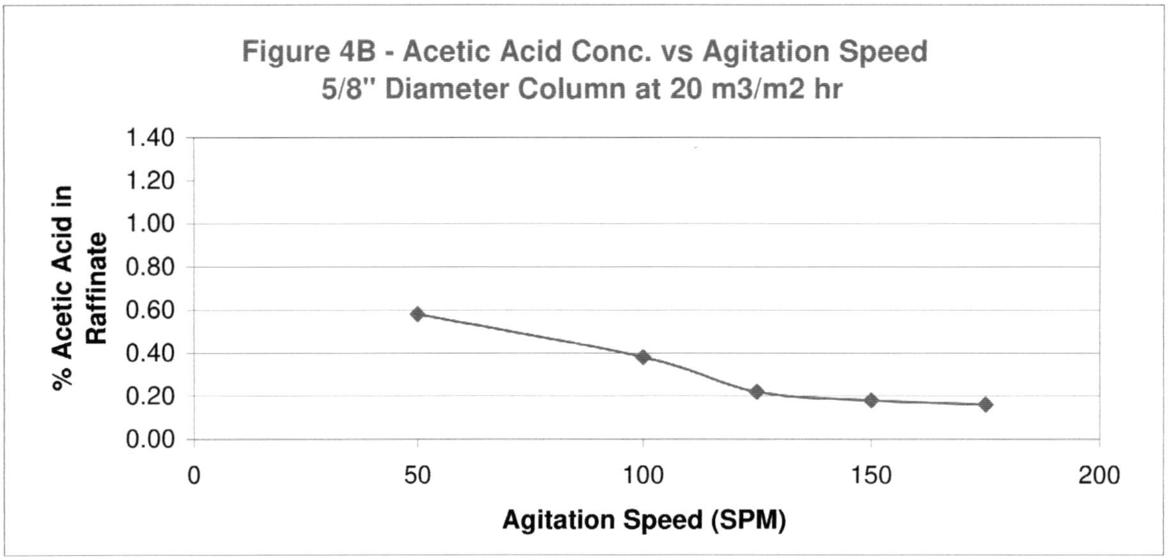
Figure 4B - Acetic Acid Conc. vs Agitation Speed
5/8" Diameter Column at 20 m3/m2 hr
% Acetic Acid in Raffinate
1.40
1.20
1.00
0.80
0.60
0.40
0.20
0.00
0
50
100
150
200
Agitation Speed (SPM)

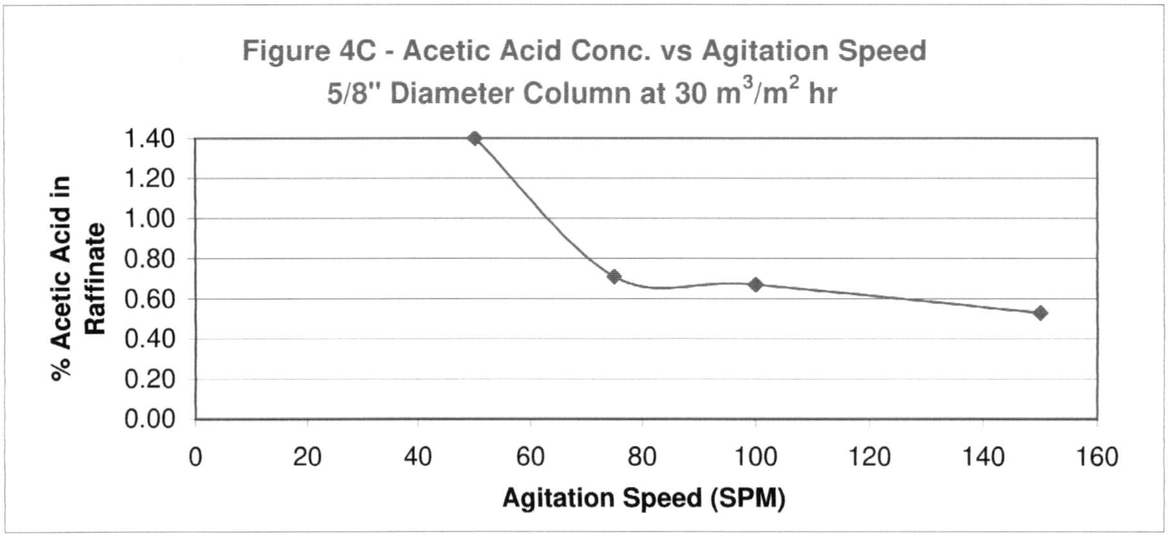
Figure 4C - Acetic Acid Conc. vs Agitation Speed
5/8" Diameter Column at 30 m3/m2 hr
% Acetic Acid in Raffinate
1.40
1.20
1.00
0.80
0.60
0.40
0.20
0.00
0
20
40
60
80
100
120
140
160
Agitation Speed (SPM)

Figure 5A

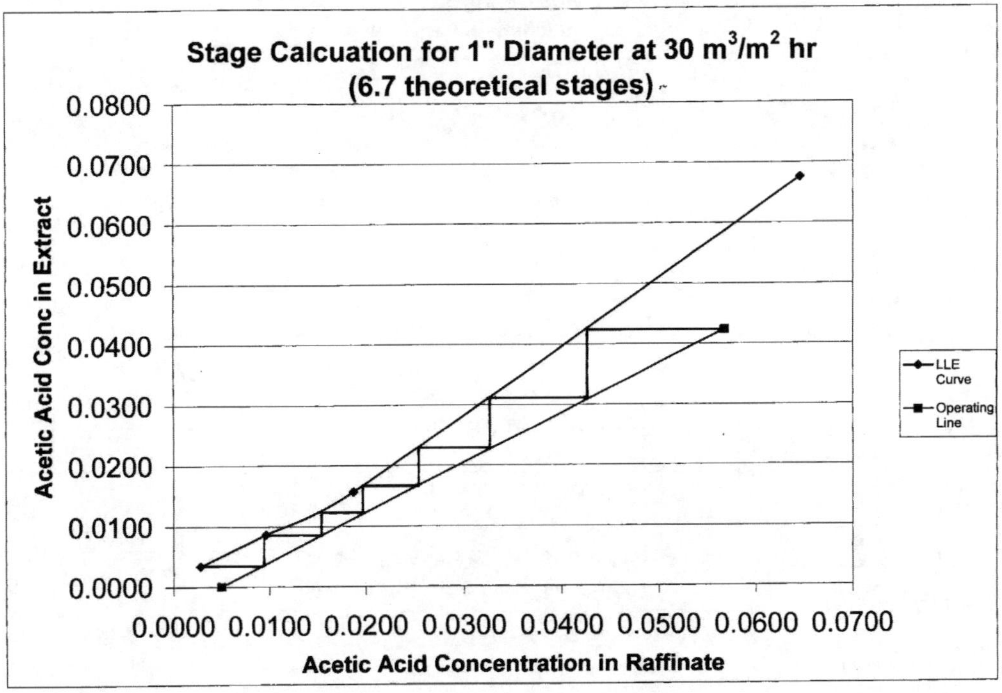

Figure 5B

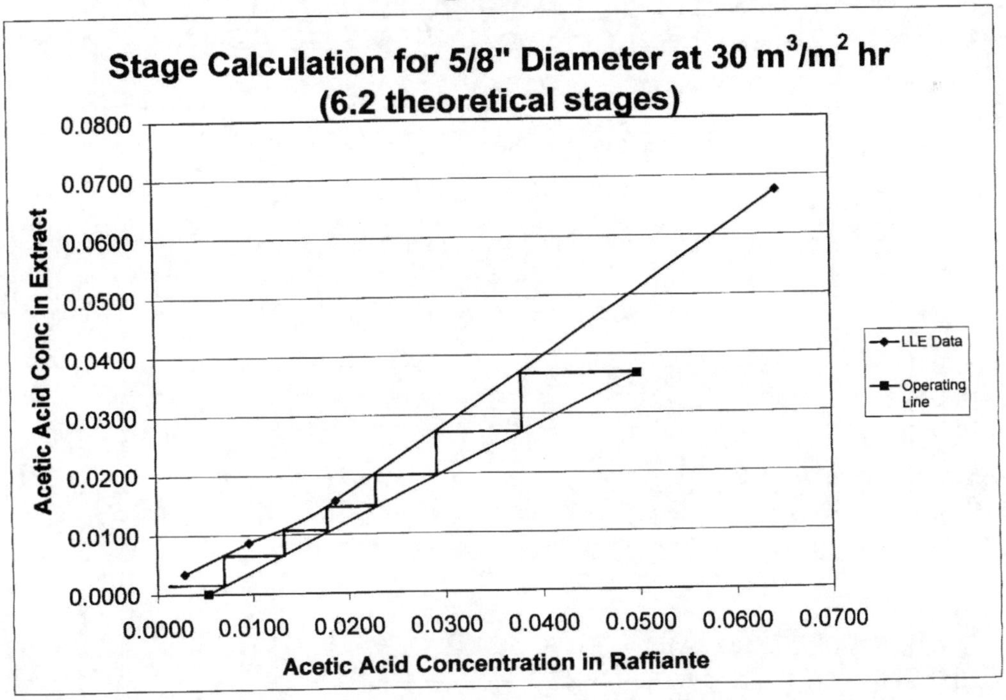

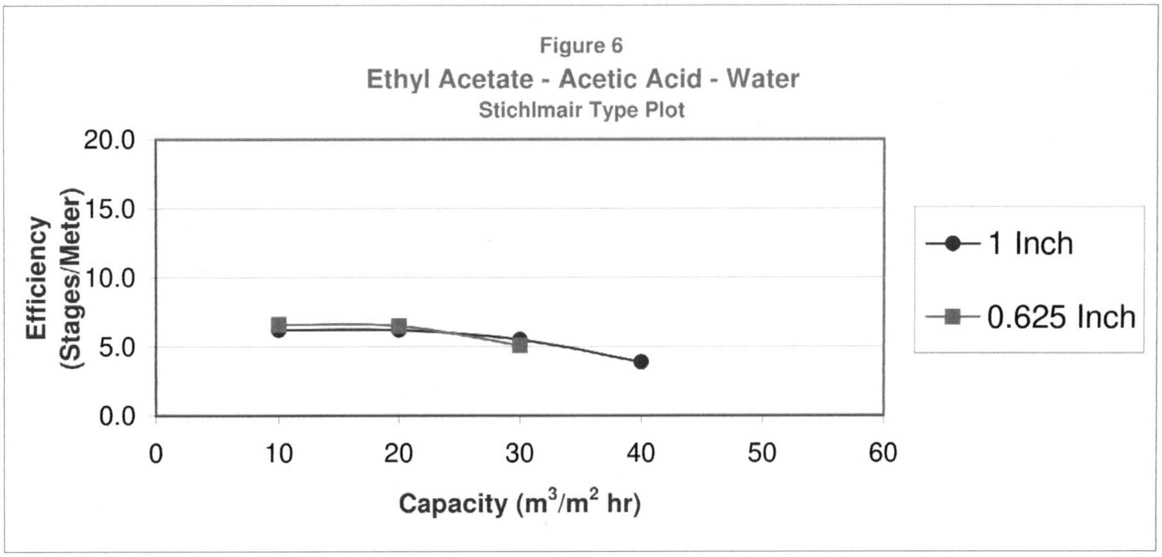

Figure 6
Ethyl Acetate - Acetic Acid - Water
Stichlmair Type Plot
Efficiency (Stages/Meter)
20.0
15.0
10.0
5.0
0.0
0
10
20
30
40
50
60
Capacity (m^3/m^2 hr)
1 Inch
0.625 Inch

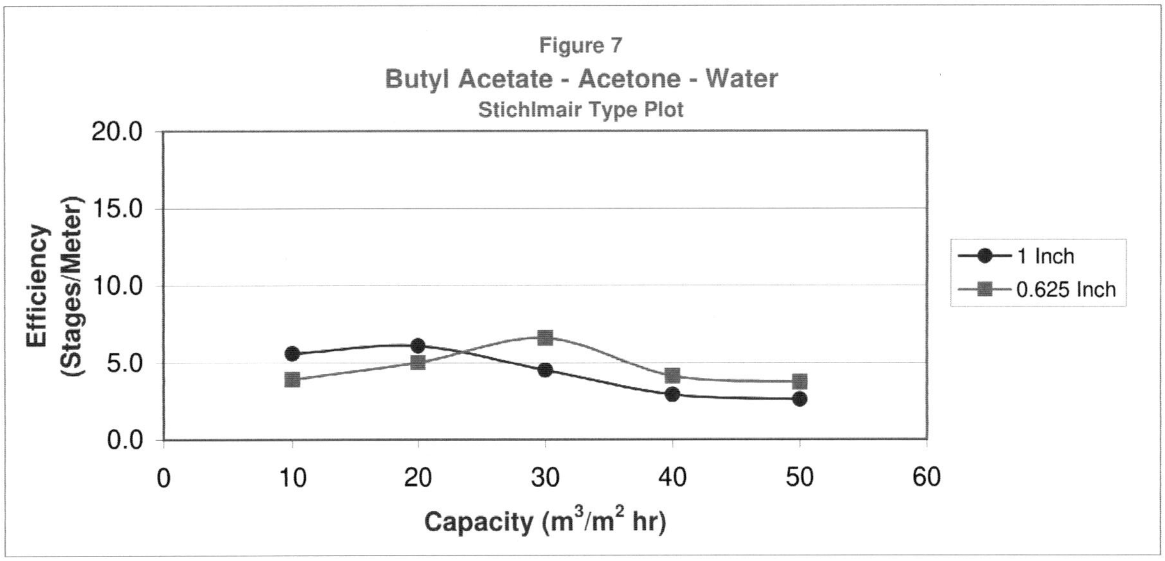

Figure 7
Butyl Acetate - Acetone - Water
Stichlmair Type Plot
Efficiency (Stages/Meter)
20.0
15.0
10.0
5.0
0.0
0
10
20
30
40
50
60
Capacity (m^3/m^2 hr)
1 Inch
0.625 Inch

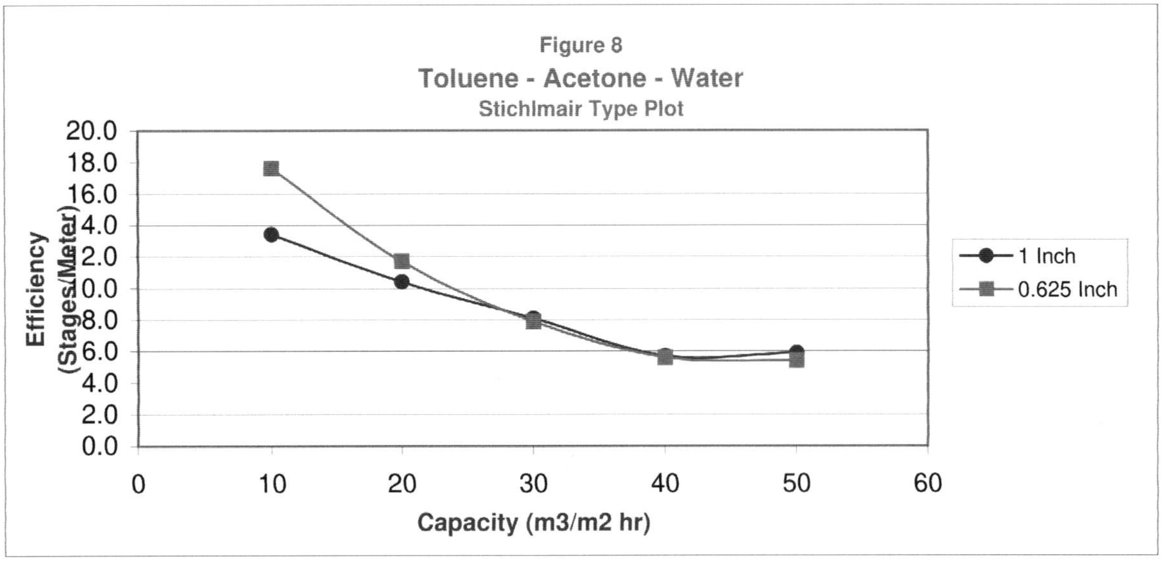

Figure 8
Toluene - Acetone - Water
Stichlmair Type Plot
Efficiency (Stages/Meter)
20.0
18.0
16.0
14.0
12.0
10.0
8.0
6.0
4.0
2.0
0.0
0
10
20
30
40
50
60
Capacity (m3/m2 hr)
1 Inch
0.625 Inch

Comparative Study of the Economic Analysis and Energy Requirements of a Non-Phosgene Process for Manufacture of Isocyanates with That of the Traditional Phosgenation Route

Nitin V. Tople, Chemical and Petroleum Engineering, University of Pittsburgh, Pittsburgh, PA

Eric J. Beckman, University of Pittsburgh, Pittsburgh, PA

Isocyanates are commonly used as monomers to make polymers, such as polyurethanes which find significant application in the manufacture of rigid and flexible foams. Polyurethanes are also used in the production of adhesives, elastomers, and coatings. Isocyanates are currently manufactured by the phosgenation route which involves reacting an amine with phosgene. This process requires handling of hazardous chemicals and is also environmentally malignant due to issues related to disposal of the hydrochloric acid, formed as a byproduct.

McGhee et al have previously worked on a route to manufacture isocyanates from primary amine, carbon dioxide, a nitrogenous base and an electrophilic dehydrating agent. We modified this route by eliminating the use of the nitrogenous base and hence reducing the salt formed. Also by maintaining either moderately high pressure of carbon dioxide or extremely low temperatures it is possible to maintain huge excess of carbon dioxide in the reaction mixture. In excess of carbon dioxide, the zwitterion intermediate of the primary amine is formed which on reaction with a dehydrating agent forms isocyanate without any salt as byproduct. We have studied the effect of different solvents, dehydrating agents and varying temperature and pressures on the overall yield of isocyanate with the aim of maximizing the isocyanate content. The energy requirements of the process at different operating conditions for the non-phosgene route are compared with the traditional phosgenation route by simulating a plant for isocyanate manufacture using ASPEN. The economic analysis for project financing and operating cost for the two routes is presented using ASPEN - Icarus Process Evaluator.

McGhee, W.D., Paster M.D., Riley D.P., Ruettimann K.W., Solodar J.A, Waldman T.E, U.S. Patent # 5,451,697, 1995

See more of this Session: Advances in Process Intensification II
See more of this Group/Topical: Process Development Division

Synthesis of Heat Exchanger Networks with Optimal Placement of Utilities

José M. Ponce-Ortega, Universidad Michoacana de San Nicolás de Hidalgo, Morelia, Mich. 58060, México
Medardo Serna-González, Universidad Michoacana de San Nicolás de Hidalgo, Morelia, Mich. 58060, México
Arturo Jiménez-Gutiérrez, Instituto Tecnológico de Celaya, Celaya, Gto. 38010, México

1. Introduction

Heat exchanger network synthesis has been widely studied because of the economic and environmental benefits such networks produce. The paper review by Furman & Sahinidis (2002) reports the techniques proposed to solve the heat exchanger network (HEN) problem. The approaches to solve the HEN synthesis problem have been classified as simultaneous and sequential. Pinch technology (Linnhoff & Hindmarsh, 1983) is the most widely used sequential approach to solve the HEN problem. The method decomposes the problem into two stages. Targets are first found for utilities, number of units and area requirements, and then the objective becomes the synthesis of a network that meets such targets. Applications of pinch concepts into some simple optimization methods have been reported, such as the transportation formulation by Cerda, Westerberg, Mason & Linnhoff (1983), which was used to predict the minimum utility requirement, and the transshipment approach by Papoulias & Grossmann (1983), in which the selection of multiples utilities was included. Simultaneous techniques take into account trade-offs between capital and energy costs, and are based on the use of mathematical programming, such as mixed-integer nonlinear programming (MINLP) methods.

Most of the methodologies reported to solve the HEN synthesis problems have considered that only one type of hot and cold utilities is available; however, in practice there are frequently several types of hot utilities (for example steam at different pressure levels) and cold utilities (for example cooling water and air cooling) that can be used. At first glance, one could assume that, whenever possible, it would be preferable to use the utility with the lower unit cost; however, the utility with the lowest cost may produce the highest capital cost because of their lower film heat transfer coefficients. In addition, most of the methodologies reported to solve the HEN synthesis problem do not optimize the location for the utilities within the network; the hot utilities are placed at the hottest side of the cold process streams, whereas the cold utilities are located at the coldest side for the hot process streams. Therefore, it is important to consider simultaneously the optimal selection of the type of utility and its location within the network.

Shenoy, Sinha and Bandyopadhyay (1998) proposed a targeting methodology to determine the optimum loads for multiple utilities in the synthesis of HENs. Their method was based on the cheapest utility principle, which states that it is better to increase the load of the cheapest utility while maintaining the loads of the relatively expensive utilities constant. Salama (2009) improved the method by Shenoy, Sinha and Bandyopadhyay (1998) by modifying the sequential placement of the multiple utilities.

From the works based on mathematical programming, the SYNHEAT model reported by Yee & Grossmann (1990) stands as one of the most notable methods. A convenient characteristic of the SYNHEAT model is that all of the constraints are linear. In the SYNHEAT model, the utilities are located at the extremes of the superstructure. Recently, Isafiade & Fraser (2008) proposed an interval-based MINLP superstructure model for the synthesis of HENs. This method is similar to the superstructure model of Yee & Grossmann (1990), but the utilities are treated as process streams, and the border temperatures of the superstructure are fixed by the supply and target temperatures of the process streams. In addition, since the model by Isafiade & Fraser (2008) treats the utilities flowrates as optimization variable, the heat balances in each stage of the superstructure yield nonlinear relationships.

It should also be noted that none of the previous methodologies that treat multiples utilities have considered isothermal streams with phase change. Recently, the paper by Ponce-Ortega, Jiménez-Gutiérrez & Grossmann (2008) has shown the importance to consider this type of streams in the HEN synthesis problems.

In this work, a new representation for the synthesis of HENs that allows for multiple utilities, along with their optimal placement within the network is proposed. The trade-offs between the capital and utility costs for the use of utilities is modeled through an MINLP problem. In addition to process streams that exchange sensible heat, the model includes isothermal process streams that exchange their latent heat. The application of the proposed model is shown through an example problem.

2. Outline of the Proposed Model

Figure 1 exemplifies the proposed superstructure for two hot and two cold process streams. This superstructure is similar to the stage-wise representation by Yee & Grossmann (1990), in which the problem is divided into stages, and all possible exchanges between a hot and a cold process stream are allowed. The temperatures for the borders of each stage are treated as optimization variables. Another important characteristic for this formulation is that isothermal mixing at the exit of each stage of the superstructure is assumed to avoid nonlinear heat balances. Notice in the representation of Figure 1 that the utilities are allowed to be placed within each stage of the superstructure. Thus, the optimal solution will reflect the best choice of the type of utility as well as its proper location within the network.

3. Model Formulation

The model formulation consists of a set of constraints for the superstructure shown in Figure 1. The following sets are used in the model formulation. *HPS* is a set that represents the hot process streams, *CPS* is a set for the cold process streams, and *ST* is a set for the number of stages in the superstructure. *HPS1* and *HPS2* are subsets for the hot process streams for the nonisothermal and isothermal streams, respectively ($HPS = HPS1 \cup HPS2$), whereas *CPS1* and *CPS2* are subsets for the nonisothermal and isothermal cold process streams ($CPS = CPS1 \cup CPS2$). *HU* is the set for the hot utilities and *CU* for the cold utilities. The indexes *i*, *j* and *k* are used to denote hot process streams, cold process streams and the stage in the superstructure, respectively; *m* is an index for the cold utilities and *n* for the hot utilities. A list of symbols can be found in a previous work (Ponce-Ortega, Jiménez-Gutiérrez & Grossmann, 2008); the list has not been included here because of space limitations. The model is described as follows.

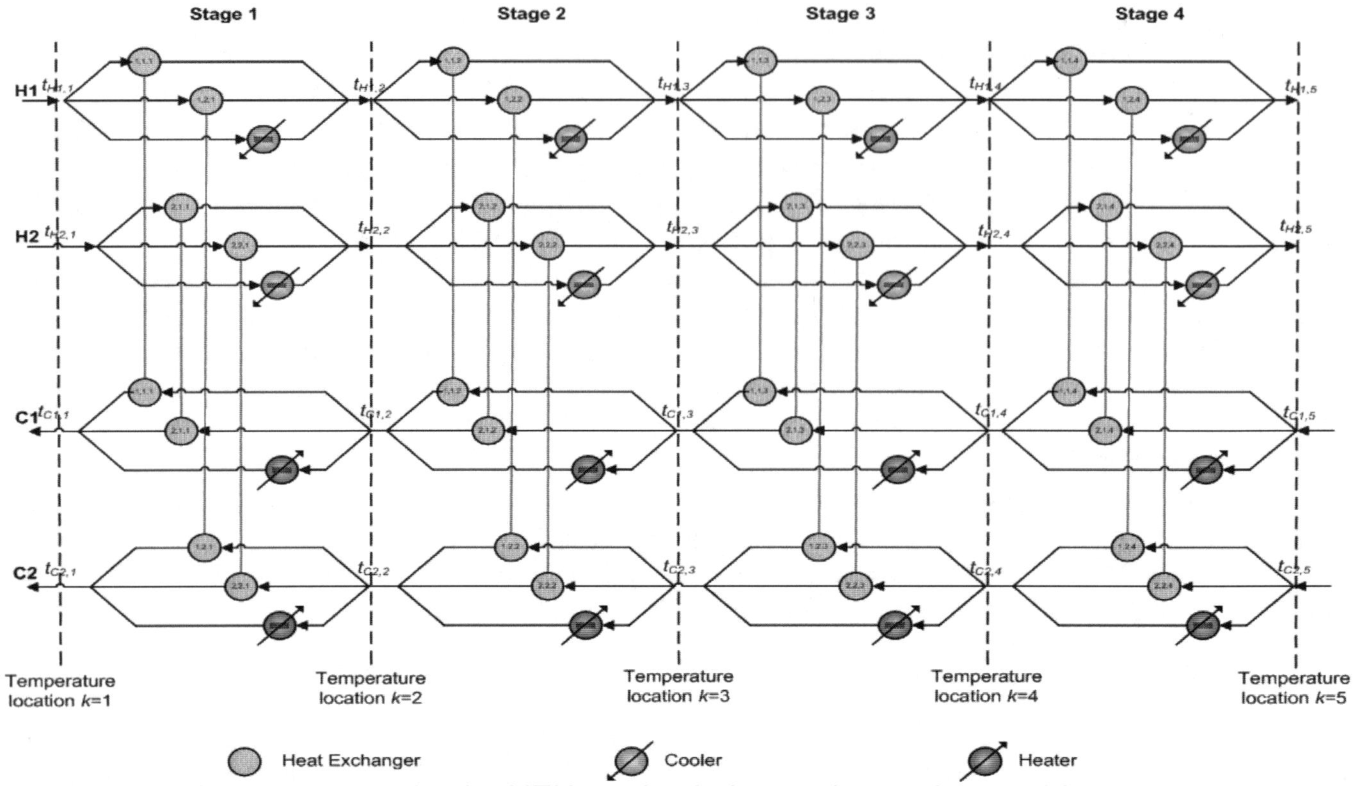

Figure 1. Superstructure for the HEN synthesis for two hot and two cold process streams

a) Overall Energy Balance for Each Process Stream

$$\left(TIN_i - TOUT_i\right)FCp_i = \sum_{k\in ST}\sum_{j\in CPS} q_{i,j,k} + \sum_{k\in ST}\sum_{m\in CU} qcu_{i,k}^m, \quad i\in HPS1 \tag{1}$$

$$F\lambda_i^{cond} = \sum_{k\in ST}\sum_{j\in CPS} q_{i,j,k} + \sum_{k\in ST}\sum_{m\in CU} qcu_{i,k}^m, \quad i\in HPS2 \tag{2}$$

$$\left(TOUT_j - TIN_j\right)FCp_j = \sum_{k\in ST}\sum_{i\in HPS} q_{i,j,k} + \sum_{k\in ST}\sum_{n\in HU} qhu_{j,k}^n, \quad j\in CPS1 \tag{3}$$

$$F\lambda_j^{evap} = \sum_{k\in ST}\sum_{i\in HPS} q_{i,j,k} + \sum_{k\in ST}\sum_{n\in HU} qhu_{j,k}^n, \quad j\in CPS2 \tag{4}$$

b) Energy Balance for Each Stage

$$\left(t_{i,k} - t_{i,k+1}\right)FCp_i = \sum_{j\in CPS} q_{i,j,k} + \sum_{m\in CU} qcu_{i,k}^m, \quad k\in ST, i\in HPS1 \tag{5}$$

$$\left(t_{j,k} - t_{j,k+1}\right)FCp_j = \sum_{i\in HPS} q_{i,j,k} + \sum_{n\in CU} qhu_{j,k}^n, \quad k\in ST,\ j\in CPS1 \tag{6}$$

c) Assignment of Inlet Temperatures to the Superstructure

$$TIN_i = t_{i,1}, \quad i\in HPS \tag{7}$$

$$TOUT_i = t_{i,NOK+1}, \quad i\in HPS \tag{8}$$

$$TIN_j = t_{j,NOK+1}, \quad j\in CPS \tag{9}$$

$$TOUT_j = t_{j,1}, \quad j\in CPS \tag{10}$$

d) Temperature Feasibility

$$t_{i,k} \geq t_{i,k+1}, \quad k\in ST, i\in HPS1 \tag{11}$$

$$t_{i,k} = TIN_i, \quad k \in ST, i \in HPS2 \tag{12}$$

$$t_{j,k} \geq t_{j,k+1}, \quad k \in ST, j \in CPS1 \tag{13}$$

$$t_{j,k} = TIN_j, \quad k \in ST, j \in CPS2 \tag{14}$$

e) Definition for the Heat Exchangers

$$q_{i,j,k} - Q_{i,j}^{\max} z_{i,j,k} \leq 0, \quad i \in HPS, j \in CPS, k \in ST \tag{15}$$

$$qcu_{i,k}^{m} - Q_i^{\max} zcu_{i,k}^{m} \leq 0, \quad i \in HPS, k \in ST, m \in CU \tag{16}$$

$$qhu_{j,k}^{n} - Q_j^{\max} zhu_{j,k}^{n} \leq 0, \quad j \in CPS, k \in ST, n \in HU \tag{17}$$

f) Temperature Differences

$$dt_{i,j,k} \leq t_{i,k} - t_{j,k} + \Delta T_{i,j}^{\max}\left(1 - z_{i,j,k}\right), \quad i \in HPS, j \in CPS, k \in ST \tag{18}$$

$$dt_{i,j,k+1} \leq t_{i,k+1} - t_{j,k+1} + \Delta T_{i,j}^{\max}\left(1 - z_{i,j,k+1}\right), \quad i \in HPS, j \in CPS, k \in ST \tag{19}$$

$$dtcu_{i,k} \leq t_{i,k} - TOUT_{cu_{i,k}} + \Delta T_{cu_i}^{\max}\left(1 - zcu_{i,k}\right), \quad i \in HPS, k \in ST \tag{20}$$

$$dtcu_{i,k+1} \leq t_{i,k+1} - TIN_{cu_{i,k}} + \Delta T_{cu_i}^{\max}\left(1 - zcu_{i,k}\right), \quad i \in HPS, k \in ST \tag{21}$$

$$dthu_{j,k} \leq TIN_{hu_{j,k}} - t_{j,k} + \Delta T_{hu_j}^{\max}\left(1 - zhu_{j,k}\right), \quad j \in CPS, k \in ST \tag{22}$$

$$dthu_{j,k+1} \leq TOUT_{hu_{j,k}} - t_{j,k+1} + \Delta T_{hu_j}^{\max}\left(1 - zhu_{j,k}\right), \quad j \in CPS, k \in ST \tag{23}$$

$$\begin{aligned}
&\text{if} \quad TIIN_i - TJIN_j < \Delta T_{MIN} \\
&\qquad \Delta T_{i,j}^{\max} = \text{abs}\left[TIIN_i \text{ - } TJIN_j\right] + \Delta T_{MIN} \\
&\text{else} \\
&\Delta T_{i,j}^{\max} = \max\left\{0, TIIN_i - TJIN_j, TJOUT_j - TIOUT_i\right\}
\end{aligned} \tag{24}$$

$$\Delta T_{\min} \geq dt_{i,j,k}, \quad i \in HPS, j \in CPS, k \in ST \tag{25}$$

$$\Delta T_{\min} \geq dtcu_{i,k}, \quad i \in HPS, k \in ST \tag{26}$$

$$\Delta T_{\min} \geq dthu_{j,k}, \quad j \in CPS, k \in ST \tag{27}$$

g) Selection of the Type of Cold Utility

For different types of cold utilities available to satisfy the heat balances, the optimization approach must select the optimal one according to the following disjunction,

$$\begin{bmatrix} & Zcu_{i,k} & \\ \begin{bmatrix} Z_{cu_{i,k}}^{1} \\ TOUT_{cu_{i,k}} = TOUT_{cu}^{1} \\ TIN_{cu_{i,k}} = TIN_{cu}^{1} \end{bmatrix} \vee \begin{bmatrix} Z_{cu_{i,k}}^{2} \\ TOUT_{cu_{i,k}} = TOUT_{cu}^{2} \\ TIN_{cu_{i,k}} = TIN_{cu}^{2} \end{bmatrix} \vee \cdots \vee \begin{bmatrix} Z_{cu_{i,k}}^{NCU} \\ TOUT_{cu_{i,k}} = TOUT_{cu}^{NCU} \\ TIN_{cu_{i,k}} = TIN_{cu}^{NCU} \end{bmatrix} \end{bmatrix} \vee \begin{bmatrix} \neg Zcu_{i,k} \\ TOUT_{cu_{i,k}} = 0 \\ TIN_{cu_{i,k}} = 0 \end{bmatrix}, \quad i \in HPS, k \in ST$$

The disjunction is modeled using the convex hull reformulation as follows,

$$zcu_{i,k} = \sum_{m \in CU} z_{cu_{i,k}}^{m}, \quad i \in HPS, k \in ST \tag{28}$$

$$TOUTcu_{i,k} = \sum_{m \in CU} tout_{cu_{i,k}}^{m}, \quad i \in HPS, \quad k \in ST \tag{29}$$

$$TINcu_{i,k} = \sum_{m \in CU} tin_{cu_{i,k}}^{m}, \quad i \in HPS, \quad k \in ST \tag{30}$$

$$tout^{m}_{cu_{i,k}} = TOUT^{m}_{cu}\, z^{m}_{cu_{i,k}}, \quad i \in HPS, \quad k \in ST, \quad m \in CU \tag{31}$$

$$tin^{m}_{cu_{i,k}} = TIN^{m}_{cu}\, z^{m}_{cu_{i,k}}, \quad i \in HPS, \quad k \in ST, \quad m \in CU \tag{32}$$

Notice that all the equations in the disjunction are linear, which aids in the numerical solution of the model.

h) Selection of the Type of Hot Utility

The following disjunction is used for the selection of the type of hot utility,

$$\begin{bmatrix} & Zhu_{j,k} & \\ \begin{bmatrix} Z^{1}_{hu_{j,k}} \\ TOUT_{hu_{j,k}} = TOUT^{1}_{hu} \\ TIN_{hu_{j,k}} = TIN^{1}_{hu} \end{bmatrix} \vee \begin{bmatrix} Z^{2}_{hu_{j,k}} \\ TOUT_{hu_{j,k}} = TOUT^{2}_{hu} \\ TIN_{hu_{j,k}} = TIN^{2}_{hu} \end{bmatrix} \vee \cdots \vee & \begin{bmatrix} Z^{NHU}_{hu_{j,k}} \\ TOUT_{hu_{j,k}} = TOUT^{NHU}_{hu} \\ TIN_{hu_{j,k}} = TIN^{NHU}_{hu} \end{bmatrix} \end{bmatrix} \vee \begin{bmatrix} \neg Zhu_{j,k} \\ TOUT_{hu_{j,k}} = 0 \\ TIN_{hu_{j,k}} = 0 \end{bmatrix}, \quad j \in CPS, k \in ST$$

The disjunction is modeled using the convex hull reformulation as follows,

$$zhu_{j,k} = \sum_{n \in HU} z^{n}_{hu_{j,k}}, \quad j \in CPS, k \in ST \tag{33}$$

$$TOUThu_{j,k} = \sum_{n \in HU} tout^{n}_{hu_{j,k}}, \quad j \in HPS, \quad k \in ST \tag{34}$$

$$TINhu_{j,k} = \sum_{n \in HU} tin^{n}_{hu_{j,k}}, \quad j \in CPS, \quad k \in ST \tag{35}$$

$$tout^{n}_{hu_{j,k}} = TOUT^{n}_{hu}\, z^{n}_{hu_{j,k}}, \quad j \in CPS, \quad k \in ST, \quad n \in HU \tag{36}$$

$$tin^{n}_{hu_{j,k}} = TIN^{n}_{hu}\, z^{n}_{hu_{j,k}}, \quad j \in CPS, \quad k \in ST, \quad n \in HU$$

i) Objective Function

The objective function is written as,

$$\begin{aligned} \min \sum_{i \in HPS} \sum_{k \in ST} \sum_{m \in CU} CCU^{m}\, qcu^{m}_{i,k} + \sum_{j \in CPS} \sum_{k \in ST} \sum_{n \in HU} CHU^{n}\, qhu^{n}_{j,k} \\ + \sum_{i \in HPS} \sum_{j \in CPS} \sum_{k \in ST} CF_{i,j}\, z_{,j,k} + \sum_{i \in HPS} \sum_{j \in CPS} \sum_{k \in ST} C_{i,j} \left\{ \frac{q_{i,j,k}\left(\frac{1}{h_{i,k}} + \frac{1}{h_{j,k}}\right)}{\left[\left(dt_{i,j,k}\right)\left(dt_{i,j,k+1}\right)\left(\frac{dt_{i,j,k} + dt_{i,j,k+1}}{2}\right) + \delta\right]^{1/3}} \right\}^{\beta} \\ + \sum_{i \in HPS} \sum_{k \in ST} CF_{i,cu}\, zcu_{i,k} + \sum_{i \in HPS} \sum_{k \in CU} C_{i,cu} \left\{ \frac{\sum_{m \in CU} qcu^{m}_{i,k}\left(\frac{1}{h_{i}} + \frac{1}{h^{m}_{cu}}\right)}{\left[\left(dtcu_{i,k}\right)\left(dtcu_{i,k+1}\right)\left(\frac{dtcu_{i,k} + dtcu_{i,k+1}}{2}\right) + \delta\right]^{1/3}} \right\}^{\beta} \\ + \sum_{j \in CPS} \sum_{k \in ST} CF_{cu,j}\, zhu_{j,k} + \sum_{j \in CPS} \sum_{k \in HU} C_{hu,j} \left\{ \frac{\sum_{n \in HU} qhu^{n}_{j,k}\left(\frac{1}{h^{n}_{hu}} + \frac{1}{h_{j}}\right)}{\left[\left(dthu_{j,k}\right)\left(dthu_{j,k+1}\right)\left(\frac{dthu_{j,k} + dthu_{j,k+1}}{2}\right) + \delta\right]^{1/3}} \right\}^{\beta} \end{aligned} \tag{37}$$

Here, CCU^m and CHU^n are the unit costs for the different cold and hot utilities. Notice that the objective function minimizes the total annual cost, which includes the operating costs for hot and cold utilities, and the investment requirements for heat transfer units. Since the optimal selection for the type of utility is needed, both unit costs and film coefficients are used. Chen's approximation (Chen, 1987) is used for the calculation of the log mean temperature differences.

In summary, the optimization model consists in the minimization of equation (37), subject to constrains given by (1) to (36). Notice that the nonlinearities appear only in the objective function.

4. Example

This example consists of two hot and one cold process streams. Three types of steam (high, medium and low pressure steam) and cooling water are available as external utilities. The stream data are reported in Table 1.

Table 1. Data for Example

stream	*TIN* [°C]	*TOUT* [°C]	*FCp* [KW K^{-1}]	*h* [KW/(m^2 K)]	Cost ($/(kW year)
H1	105	25	10	0.5	-
H2	185	35	5	0.5	-
HPS[1]	210	209	-	5.0	160
MPS[2]	160	159	-	5.0	110
LPS[3]	130	129	-	5.0	50
C1	25	185	7.5	0.5	-
CW[4]	5	6	-	2.6	10

Figure 2 shows the network obtained with the proposed approach. Notice that three heat exchangers are used for energy integration. Two coolers that use cooling water are required for the hot process streams *H1* and *H2*, and two heaters for the cold process stream *C1* (one using high pressure steam and the other using low pressure steam) are needed. Notice that one of the heaters (low pressure steam) is located between two exchangers for process streams, situation that can not be reproduced by the original model by Yee & Grosmmann (1990). The total heat transfer area for the 7 units of the network is 184.2 m^2, which yields a total capital cost of $43,922/year. The total utility cost is $53,157/year, and the total annual cost is $97,079/year. Table 2 summarizes the comparative results obtained in this work with the ones reported by Shenoy, Sinha & Bandyopadhyay (1998) and Isafiade & Fraser (2008) for this problem. Notice that of the three structures, the network reported in Figure 2 is the one with the highest consumption of low pressure steam and the one with lower heat transfer units. The total annual cost of the network obtained with the proposed method is 1.22% and 0.14% cheaper with respect to the solutions reported by Shenoy, Sinha & Bandyopadhyay (1998) and Isafiade & Fraser (2008), respectively.

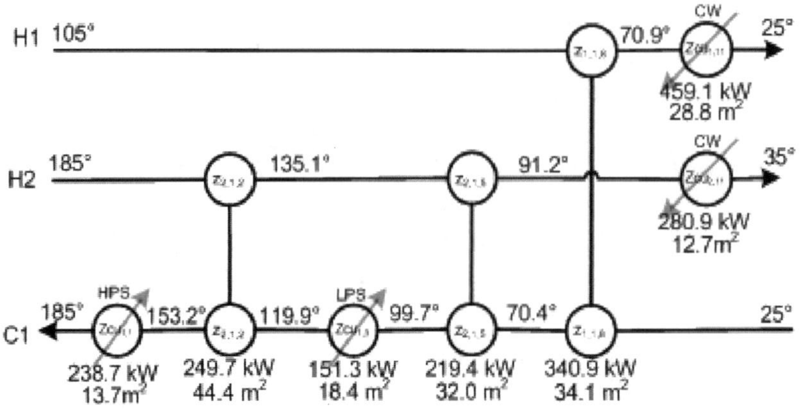

Figure 2. Optimal network configuration for Example

Table 2. Results comparison for Example

Solution Method	This paper	Shenoy, Sinha & Bandyopadhyay (1998)	Isafiade & Fraser (2008)
HPS load [kW]	238.7	203	244.6
MPS load [kW]	0	53	1
LPS load [kW]	151.3	119.5	143.7
CW load [kW]	740.0	725.5	739.3
Number of units	7	9	9
TAC [$/year]	97,079	98,263	97,211

5. Conclusions

This paper has presented an MINLP model for the optimal HEN synthesis involving different types of utilities, considering simultaneously the capital cost of heat exchangers and the operational cost of utilities. This new formulation allows the treatment of multiple utilities and their optimal location within the network, which may differ from their placement at the end of the superstructure as assumed by most of the previously reported methodologies. The optimal selection of the type of utility is modeled through a disjunctive formulation that is reformulated as an MINLP problem using the convex hull technique. The model is able to handle isothermal and nonisothermal process streams. In addition, all the constraints in the model presented are linear, with nonlinear terms appearing only in the objective function. Local optimization solvers such as DICOPT are therefore quite effective in solving the proposed model. The solution to an example taken from the literature has shown how this formulation improves the solution previously reported with other HEN synthesis models.

References

Cerda, J., Westerberg, A.W., Mason, D., & Linnhoff, B. (1983) Minimum utility usage in heat exchanger network synthesis – A transportation problem. *Chemical Engineering Science. 38* (3), 373-387.

Chen, J.J.J. (1987). Letter to the editor: comments on improvement on a replacement for the logarithmic mean. *Chemical Engineering Science, 42* (10), 2488-2489.

Furman, K.C., & Sahinidis, N.V. (2002). A critical review and annotated bibliography for heat exchanger network synthesis in the 20th century. *Industrial and Engineering Chemistry research. 41* (10), 2335-2370.

Isafiade, A.J., & Fraser, D.M. (2008). Interval-based MINLP superstructure synthesis of heat exchange networks. *Chemical Engineering Research & Design. 86* (A3), 245-257.

Linnhoff, B., & Hindmarsh, E. (1983). The pinch design method for head exchanger networks. *Chemical Engineering Science. 38* (5), 745-763.

Papoulias, S.A., & Grossmann, I.E. (1983). A structural optimization approach in process synthesis. 1. Utility systems. *Computers & Chemical Engineering, 7* (6), 695-706.

Ponce-Ortega, J.M., Jiménez-Gutiérrez, A., & Grossmann, I.E. (2008). Optimal synthesis of heat exchanger networks involving isothermal streams. *Computers & Chemical Engineering. 32* (8), 1918-1942.

Salama, A.I.A. (2009). Optimal assignment of multiple utilities in heat exchange networks. *Applied Thermal Engineering. 29* (13), 2633-2642.

Shenoy, U.V., Sinha, A. & Bondyopadhyay, S. (1998). Multiple utilities targeting for heat exchanger networks. *Chemical Engineering Research & Design. 76* (Part A), 259-272.

Yee, T.F., & Grossmann, I.E. (1990). Simultaneous optimization model for heat integration –II Heat exchanger network synthesis. *Computers & Chemical Engineering. 14* (10), 1165-1184.

A Method for Pressure Drop Targeting of Heat Exchanger Networks

Medardo Serna-González, José María Ponce-Ortega, Oscar Burgara-Montero
Universidad Michoacana de San Nicolás de Hidalgo, Department of Chemical Engineering, Morelia, México.

Abstract

This paper presents a method for targeting optimum pressure drops of streams in heat exchanger networks, which combines insights from pinch technology with mathematical programming. For that purpose, an optimization model for stream contact areas, total network area, utility consumption and stream pressure drops is developed in terms of unknown film heat-transfer coefficients. The Kern method is used to relate pressure drops, film heat-transfer coefficients and contact areas. For each energy recovery level, the stream contact areas are obtained using the *UA*-values of the individual matches from the spaghetti design. The model is formulated so as to minimize the total annual cost, which includes the capital costs of heat exchangers and pumping devices, the utilities costs and the electricity costs to run the pumping equipment. This leads to a problem which can be solved as a non-linear optimization problem. By this method, the optimal pressure drops and film heat-transfer coefficients can be calculated, and the targeting stage can remain simple. Thus, no assumed heat-transfer coefficients or fixed allowable pressure drops need to be used, as is the case with previous targeting methods based on pinch technology.

Key words: heat exchanger networks, targets, pressure drops, pumping cost.

Introduction

Heat recovery and, therefore, energy savings in the process industries require heat exchanger networks (HENs). A HEN is an interconnected arrangement of heat exchangers, mixers and splitters, in which cold and hot process streams interchange internal process heat as complete as possible for decreasing utility consumption. The objective of the HEN design problem for a given set of process streams, utility streams and cost data is to find the optimal network structure along with the optimal sizing of their heat transfer units. Typically, total annual cost has been used as an objective function for obtaining an optimum solution for the HEN design problem. Total annual cost includes the annual investment of the heat exchangers and also the annual operating costs of auxiliary utilities.

The two most common approaches that have been developed for optimal synthesis of a HEN are pinch technology and mathematical programming methods. Pinch technology is based on heuristics and thermodynamic principles (Linnhoff and Flower, 1978; Linnhoff and Hindmarsh, 1983; Linnhoff and Ahmad, 1990), while mathematical programming approach (Floudas et al., 1986; Yee and Grossmann, 1991) formulates and solves the HEN synthesis problem as a constrained optimization problem. Thorough reviews of the principles, advances and applications of both these approaches have been presented in the literature by Gundersen and Naess (1988), Jezowski (1994a, 1994b), Shenoy (1995), and Furman and Sahinides (2002). Although the mathematical programming methods have a more rigorous approach of obtaining an optimal HEN design while satisfying stipulated constraints, the pinch technology is currently more used because it allows the design engineer to incorporate real plant situations easily for industrial scale problems.

Pinch technology decomposes the HEN design problem into a sequential procedure. The first stage is to generate economic trade-offs between capital and operating costs ahead of design to select the optimal value for the minimum allowed approach temperature, ΔT_{min}. Then, in the second stage, the pinch design method is applied to develop a network based on the targets generated in the first stage. Targeting and synthesis stages constitute the conceptual design phase of pinch technology. The conceptual design is followed by the detailed design. In this stage, the detailed heat-exchanger designs based on the stream pressure drops are performed. Until recently, however, targeting and synthesis stages had been applied assuming constant film heat-transfer coefficients for all streams, which can lead to nonoptimal or infeasible networks due to inconsistencies with detailed designs of the heat exchangers at the last stage.

To overcome this limitation of pinch technology, Polley and Panjeh Shahi (1991) developed the first systematic procedure for HEN targeting considering pressure drop effects. The original area targeting algorithm was modified to include an expression that relates the stream pressure drop to the heat exchanger area and the stream heat transfer coefficient. Values of the allowable pressure drops of streams, fluid properties, volumetric flow rate and hydraulic diameters of the exchangers are specified to calculate the film heat-transfer coefficients of streams and network area. The calculated heat transfer coefficients are then used in the synthesis stage. This work improves the consistency between conceptual design and detailed design phases for given allowable pressure drops of streams. However, for grassroots designs, the pressure drops of streams are usually unknown and can be treated as part of the degrees of freedom of the HEN design problem. Therefore, it is convenient to consider the optimization of pressure drops instead of specifying fixed allowable pressure drops in the targeting stage.

As a step forward to the optimal design of HENs based on pinch technology, a mathematical model that treats as optimization variables the stream pressure drops at the targeting stage is proposed in this work. Equations based on the Kern method that relate the pressure drops, the film heat transfer coefficients and the heat exchanger geometry for heat exchange matches are included in this formulation. The objective function minimizes the total annual cost (TAC) that equals the sum of the utility consumption cost, the capital investment for the heat exchangers and pumping devices, and the pumping cost. An example from the literature that includes the detailed design for each heat exchanger of the network is used to test the method developed. The results confirm that this approach provides the consistency of heat exchanger design information through conceptual design and detailed design phases.

Problem statement

The problem addressed in this paper can be stated as follows:

Given:

- A set of hot process streams (to be cooled) and a set of cold process streams (to be heated) for heat exchange with their specified inlet and target temperatures, flow rates and physical properties (heat capacity, density, viscosity, thermal conductivity, and fouling factor);
- A single heating utility and a single cooling utility to satisfy the energy requirements with their inlet and target temperatures as well as their physical properties;
- Operating and capital cost data involved in the network installation and operation.

The objective is to develop a method based on pinch technology for obtaining the minimum total annual cost target for HEN synthesis problems, including the target for the optimum pressure drops or power cost in addition to traditional HEN design targets for minimum utilities, area and number of units. To achieve the above objective, we develop a mathematical model that is based on the *spaghetti* design model provided by the composite enthalpy-temperature curves. As will be shown, the underlying mathematical formulation corresponds to a non-linear optimization problem, where film heat transfer coefficients of the process streams are continuous variables to be selected by an optimization procedure.

The basic assumptions that will be made for this problem are:

1. The physical properties of the process streams are temperature independent.
2. The inlet temperatures of the utilities are such that they cover the feasible temperature range of the process streams that are considered.
3. The heat exchange between the hot and the cold composite curves is vertical across the whole enthalpy range.
4. There are not matching constraints between specific hot and cold streams.
5. The film heat-transfer coefficient of a stream in a network is constant, match independent.
6. Consideration is only given to heat exchange in a single phase.
7. Only shell-and-tube heat exchangers are considered.

Model formulation

The composite curves are one of the more important concepts available through the use of Pinch Technology. As is well known, the cold and hot composite curves of a process result by simple addition of the cold and hot streams in a temperature-enthalpy diagram. The minimum vertical separation between the composites curves is called the "pinch point" and corresponds to the minimum temperature approach ΔT_{min} for heat exchange. For a given ΔT_{min}, the information obtained from the composite curves involves the minimum utility consumption (Hohmann, 1971; Linnhoff and Flower, 1978) and the minimum network heat transfer area for specified utilities (Hohmann, 1971; Linnhoff and Ahmad, 1990). These two objectives impose targets that are to be achieved in the synthesis of HENs. For a given value of ΔT_{min}, the target for a minimum number of units for specified utilities also can be estimated before synthesis by using a simplified form of Euler's graph theory (Hohmann, 1971; Linnhoff and Flower, 1978). It should be noted that the minimum number of units is independent of area and the minimum network area is independent of the number of units.

The target for HEN capital cost can be predicted by combining the targets for the minimum network area and the number of units together with the heat exchanger cost law equations (Linnhoff and Ahmad, 1990). The energy cost target and the capital cost target can then be combined to obtain the minimum total annual cost (TAC) target for HENs. This provides an optimum ΔT_{min} prior to determining the HEN structure (Linnhoff and Ahmad, 1989; 1990). However, until now, there is no way of targeting the cost optimum pressure drops of streams for HENs. This is because the existing targeting procedures consider constant values of film heat-transfer coefficients or fixed stream pressure drops of process streams, which are given as problem data based on the designer experience. In this section, therefore, we develop a model for considering the optimization of pressure drops since the targeting stage.

Objective function. The objective function is given in the form

$$min\ TAC = A_f\, CC + C_U + C_P \qquad (1)$$

where A_f is the annualized factor for investment, CC is the capital cost for heat exchange units, C_U is the hot and cold utilities costs, and C_P is the power cost.

Capital cost. Assuming an equal distribution of area among all heat exchangers in the HEN, the network capital cost is given as

$$C_{CAP} = N_{u,mer} [a + b(A_{min}/N_{u,mer})^c] \quad (2)$$

where $N_{u,mer}$ is the minimum number of units for a maximum energy recovery network, A_{min} is the area target, and *a*, *b*, and *c* are cost law coefficients that depend on the type of heat exchanger, construction materials, and pressure rating. Values for *c* are typically between 0.6 and 0.9, which reflect the economies of scale commonly observed in chemical process equipment. It should be noted that for these fractional values of *c*, the network capital cost is an exponential function of the overall area target.

Utility cost. The annual utility cost of a network is given by

$$C_U = H_Y (C_H Q_{Hmin} + C_C Q_{Cmin}) \quad (3)$$

Here Q_{Hmin} (Q_{Cmin}) is the minimum hot (cold) utility target, C_H (C_C) is the unit cost of hot (cold) utility, and H_Y represents the hours of operation of network per annum.

Power cost. When fluid streams go through heat exchangers, there is a loss of pressure caused by internal friction and expansion and contraction of the fluids at the inlet and outlet of the heat exchanger. This pressure drop has to be compensated by supplying mechanical energy with external pumps. The required power, for pumping the fluid streams, is given by the product $Q_i \Delta P_i$, where Q_i is the volumetric flow rate of the process stream *i* and ΔP_i is the pressure drop of the same stream. Pressure drop for a hot stream *i* may be evaluated by the following equation (Polley and Panjeh Shahi, 2001).

$$\Delta P_i = K_i A_{ci}\, h_{c,i}^{\,m} \qquad \text{for } I \in I \quad (4)$$

where, for the stream *i*, A_{ci} is the contact area of the heat exchanger, $h_{c,i}$ is the clean film heat-transfer coefficient, and exponent *m* and constant K_i depend on the physical properties of the stream and the geometry of the heat exchanger. The contact area is function of the dirt film heat-transfer coefficients, as shown in Eq. (5) for a single exchanger that matches hot stream 1 and cold stream 2.

$$A_{c1} = A_{c2} = \frac{q}{\Delta T_{LM}}\frac{1}{U_{12}} = \frac{q}{\Delta T_{LM}}\left[\frac{1}{h_1} + \frac{1}{h_2}\right] \quad (5)$$

Clean and dirt heat-transfer coefficients are related by Eq. (6).

$$h_{ci} = \frac{1}{\left[\frac{1}{f_i h_i} - R_{di}\right]} \qquad \text{for } i \in I \quad (6)$$

where R_{di} is the fouling factor for stream *i*, f_i is a constant factor equals to one when the stream goes through the shell and equals to D_t/D_{ti} (external and internal tube diameter) when the stream goes inside the tubes.

The power cost for stream *i* may be expressed as follows:

$$C_{POW,i} = \frac{H_Y C_{P,i} Q_i K_i A_{ci}}{\left[\frac{1}{f_i h_i} - R_{di}\right]^m} \qquad \text{for } i \in I \quad (7)$$

where $C_{p,i}$ is the unit cost of power used by stream *i*, and A_{ci} is the contact area for the same stream. The expressions for cold streams may be obtained substituting the subscript *i* by *j* and exponent *m* by *n*.

Contact areas. The contact areas can not be directly determined from balanced composite curves of the system, because neither film heat-transfer coefficients nor pressure drops of streams are specified. Therefore, it is necessary to develop a relationship between contact areas and film heat-transfer coefficients. The balanced composite curves provide such relationship in terms of the UA-values of the individual matches that constitute the *spaghetti* design.

For a given value of ΔT_{min}, the first step to develop the *spaghetti* design of a process is to divide the balanced composite curves into k = 1, ..., K enthalpy intervals defined wherever a change in slope occurs in either composite profile (Linnhoff and Ahmad, 1990). The *spaghetti* design assumes that each hot stream is integrated with all cold streams within each enthalpy interval in order to satisfy vertical heat transfer. This generates a structure featuring parallel stream splitting and isothermal mixing. Thus, if there are I hot streams and J cold streams in a given enthalpy interval k, then $I \cdot J$ is the number of matches in the *spaghetti* design. For each match between streams i and j in the enthalpy interval k of the *spaghetti* design, Eq. (8) can be used to calculate the UA-value (Shenoy, 1995):

$$UA_{i,j,k} = \frac{q_{i,j,k}}{F_{Tk} \Delta T_{MLk}} \qquad \text{for } i \in I,\ j \in J,\ k \in K \tag{8}$$

where $q_{i,j,k}$ is the amount of heat transferred from hot stream i to cold stream j in enthalpy interval k, F_{Tk} is the correction factor for the logarithmic mean temperature difference in enthalpy interval k, and ΔT_{MLk} is the logarithmic mean temperature difference for any pair of streams in enthalpy interval k. Since the temperatures of all the hot and cold streams follow the same profiles, ΔT_{ML} (F_T) is the same for all matches in each enthalpy interval.

The heat duty $q_{i,j,k}$ is given by

$$q_{i,j,k} = \frac{CP_{i,k} CP_{j,k} \Delta T_{i,k}}{\sum_j CP_{j,k}} = \frac{CP_{i,k} CP_{j,k} \Delta T_{j,k}}{\sum_i CP_{i,k}} \qquad \text{for } i \in I,\ j \in J,\ k \in K \tag{9}$$

Here, for each enthalpy interval k, $CP_{i,k}$ is the heat capacity flow rate of hot stream i, $CP_{j,k}$ is the heat capacity flow rate of cold stream j, and $\Delta T_{i,k}$ and $\Delta T_{j,k}$ are the temperature changes of any hot and cold streams i and j. It should be noted that the use of ratios of CPs of the streams for each interval k of the composite curves is needed to keep the enthalpy balance of the system.

The total UA-value for each match (i,j) is expressed as

$$UA_{i,j} = \sum_{k=1}^{K} UA_{i,j,k} \qquad \text{for } i \in I,\ j \in J \tag{10}$$

With these $UA_{i,j}$ values and assuming that the film heat-transfer coefficients of streams are not all identical but are match independent, the contact areas may be calculated using Eq. (11a) for hot streams and (11b) for cold streams.

$$A_{ci} = \sum_{j=1}^{J} \left(\frac{1}{h_i} + \frac{1}{h_j} \right) \left(UA_{i,j} \right) \qquad \text{for } i \in I \tag{11a}$$

$$A_{cj} = \sum_{i=1}^{I} \left(\frac{1}{h_i} + \frac{1}{h_j} \right) \left(UA_{i,j} \right) \qquad \text{for } j \in J \tag{11b}$$

where A_{ci} is the contact area of the hot stream i, and A_{cj} is the contact area of the cold stream j.

Eqs. (4) and (11a, b) assume that all UA-values involving hot (cold) stream i (j) are lumped into a single exchanger that has an area equal to A_{ci} (A_{cj}). As a result, A_{ci} (A_{cj}) is contribution to network area target from hot (cold) stream i (j). Thus, the total area target of the network can be expressed as the sum of the of the area targets A_{ci} (A_{cj}) of the hot (cold) streams,

$$A_{min} = \sum_i A_{ci} = \sum_j A_{cj} \tag{12}$$

Objective function in terms of heat transfer coefficients. The objective function in terms of the heat transfer coefficients is found from the combination of Eqs. (1), (2), (3), and (7) to (11). Thus, for specified ΔT_{min}, the objective function is rewritten as follows:

$$min\,TAC = A_f N_{u,mer}\left[a + b\left(\frac{\sum_{i=1}^{I}\left[\sum_{j=1}^{J}\left(\frac{1}{h_i}+\frac{1}{h_j}\right)(UA_{ij})\right]}{N_{u,mer}}\right)^c\right]$$

$$+\sum_{i=1}^{I}\left[\frac{H_Y C_{P,i} Q_i K_i}{\left(\frac{1}{f_j h_j}-R_{di}\right)^m}\right]\left[\sum_{j=1}^{J}\left(\frac{1}{h_i}+\frac{1}{h_j}\right)(UA_{ij})\right]+H_Y C_H Q_{H\,min}$$

$$+\sum_{j=1}^{J}\left[\frac{H_Y C_{P,j} Q_j K_j}{\left(\frac{1}{f_j h_j}-R_{dj}\right)^n}\right]\left[\sum_{i=1}^{I}\left(\frac{1}{h_i}+\frac{1}{h_j}\right)(UA_{ij})\right]+H_Y C_C Q_{C\,min} \tag{13}$$

subject to:

(a) Nonnegative heat-transfer coefficients

$$h_i > 0 \qquad \text{for } i \in I$$

$$h_j > 0 \qquad \text{for } j \in J$$

As can be seen, minimizing total annual cost is a non-linear programming (NLP) problem. Note that the independent variables over which the minimization is performed are the film heat-transfer coefficients of the streams. For a given value of ΔT_{min}, the other variables can be determined before the NLP is solved. The minimum utility heating Q_{Hmin} and cooling Q_{Cmin} requirements are calculated from the problem table algorithm by Linnhoff and Flower (1978). This algorithm also provides the pinch location. The target for minimum number of exchanger units $N_{u,mer}$ is obtained by applying the Euler's equation (Linnhoff and Flower, 1978). If the HEN design problem is pinched, the unit targeting procedure can be applied to the stream populations above and below the pinch; in this case, the unit target is obtained by adding the two quantities. The UA-values ($UA_{i,j}$) of stream matches are calculated from Eqs. (8) to (10). Then, it is possible to solve the NLP for obtaining the optimum film heat-transfer coefficient of each stream. The optimum h-values can be used to determine the HEN targets for the minimum area, the capital and power costs, as well as the optimum pressure drops of the streams. On the other hand, knowing the heat transfer coefficients, the synthesis stage can be carried out using the method by Linnhoff and Hindmarch (1983) to establish the HEN structure.

Optimum ΔT_{min} prior to design. By varying the value of ΔT_{min} used for utility targeting and recalculating the unit, area and pressure drop targets, the trade-off between capital and operating costs can be predicted to find the cost-optimal value of ΔT_{min} for HEN design.

Results

The proposed procedure for predicting the minimum TAC of HENs, including the target for optimum pressure drops of streams, has been applied to solve a literature example recently considered by Mizutani et al. (2003) and Ponce-Ortega et al. (2007). The product $H_Y C_H$ was assumed as 60 ($/kW year), the product $H_Y C_H$ was taken as 6 ($/kW year), and the annualized factor for the capital cost as 0.2983/year. The values of *a, b,* and *c* were taken as 1000, 60 and 0.6, respectively. The solver CONOPT included in the GAMS optimization package (Brooke et al., 2006) was used to solve the proposed NLP problem.

The example has two hot and two cold process streams, along with one hot and one cold utility. The stream data are given in Table 1. A constant overall heat transfer coefficient for utility units is assumed (444 W/m^2 K). Physical properties, which are assumed to be constant, are given in Table 2. A total annual cost target is required at a ΔT_{min} = 10 K.

Table 1. Stream data for Example

Stream	M (kg/s)	TIN (K)	TOUT(K)
H1	8.15	368	348
H2	81.5	353	348
C1	16.3	303	363
C2	20.4	333	343
HU		500	499
CU		300	320

Table 2. Fluid physical properties for Example

Property	Value
μ [kg/m s]	2.4X10^{-4}
ρ [kg/m^3]	634
Cp [J/kg K]	2454
k [W/m K]	0.114

With the value of ΔT_{min} (10 K) specified for utility targeting, the heating and cooling targets are 1500.621 and 0 kW, respectively. The NLP model for this problem has 4 continuous variables. The optimal solution is shown in Table 3. The targeting model provides a minimum TAC of $96,818.57/year and an area target of 157 m^2.

Table 3. Results for Example

Stream	Film heat-transfer coefficient (kW/m^2 K)	Pressure drop (Pa)
H1	728	2854
H2	828	767
C1	877	5889
C2	500	351

This example was also solved by Mizutani et al. (2003) using MINLP techniques and by Ponce-Ortega et al. (2007) using genetic algorithms, considering the detailed design for the heat exchangers with the Bell-Delaware method. As shown in Table 4, results are very satisfactory as the target for minimum TAC is within an insignificant percent of the minimum cost detailed designs. However, it should be stressed that, as opposed to the detailed design approaches reported by Mizutani et al. (2003) and Ponce-Ortega et al. (2007), the proposed NLP model is very simple because it only involves 4 continuous variables.

Table 4. Comparison of results for Example

	This work	Mizutani et al. (2009)	Ponce et al. (2007)
TAC ($/year)	96,818	95,853	95,699
A_{min} (m^2)	157	111	92

Conclusions

For new designs, the pressure drops of streams are usually unknown and can be treated as part of the degrees of freedom of the HEN problem. Therefore, it is convenient to consider the optimization of pressure drops from the targeting stage. This paper presents a mathematical model to set pressure drops or power cost targets in addition to traditional HEN design targets for the minimum utilities, area and number of units. For a given value of ΔT_{min}, these targets are dependent on the economic data for the process but are independent on the network structure. Thus, this model allows the consideration of the trade-offs between all above HEN targets to obtain an optimum value of ΔT_{min} before an HEN is synthesized.

Ponce-Ortega et al. (2007) develop a method for simultaneous HEN synthesis and detailed design of the heat exchangers in the network that uses genetic algorithms. In this work, the example shows that the proposed targeting model can approach within a few percent the total annual cost targets with the optimum TAC of detailed HEN obtained by extremely rigorous methods like the one developed by Ponce-Ortega et al. (2007). Thus, the film heat-transfer coefficients and pressure drops of streams given by the solution of the NLP provide a good starting point for rigorous methods for the synthesis of detailed HENs.

However, despite the excellent targets for minimum TAC given by the NLP model, there are at least two problems with the targeting procedure. First, it cannot obtain capital cost and operation cost targets for networks comprising different exchanger specifications. Second, it cannot consider the match dependency of the film heat-transfer coefficients of streams. Now, the proposed targeting approach is being extended to consider such cases.

Acknowledgement

The authors wish to thank to CONACYT (Project I32934-U) and to the Council for Scientific Research of the Universidad Michoacana de San Nicolás de Hidalgo (Project 20.1) for the financial supports granted for the accomplishment of this work.

Nomenclature

A = heat exchanger area
A_f = annualized factor for capital costs
A_{min} = network area target
a,b,c = coefficients in capital cost laws for heat exchangers
CC = network capital cost
C_C = unit cost of cold utility
C_H = unit cost of hot utility
Cp = fluid specific heat at constant pressure and average temperature

C_{POW} = power cost
C_U = network cost of hot and cold utilities
CU = cold utility
F_T = correction factor for logarithm mean temperature difference
h = film heat-transfer coefficient
H_Y = annual plant operation time
HU = hot utility
k = thermal conductivity
m = exponent for shell-side heat transfer coefficient in pressure drop relationship
M = fluid mass flow rate
n = exponent for tube-side heat transfer coefficient in pressure drop relationship
$N_{u,mer}$ = target for minimum number of heat exchange units
q = heat duty
Q_H = hot utility consumption
Q_C = cold utility consumption
R_d = fouling factor
TAC = total annual cost
TIN = inlet temperature
$TOUT$ = outlet temperature
U = overall heat transfer coefficient

Greek symbols
ΔT_{min} = minimum temperature approach
ΔT_{LM} = log mean temperature difference
ΔP = pressure drop for process stream
μ = viscosity of fluid stream
η = efficiency of pump
ρ = density of fluid stream

Subscripts
i = hot process stream
j = cold process stream
k = enthalpy interval

Literature Cited

Brooke, A., Kendrick, D., and Meeraus, A. GAMS User's Guide, The Scientific Press, USA, **2006**.

Floudas, C.A., Ciric, A. R., Grossmann, I.E. Automatic synthesis of optimum heat exchanger network configurations. *American Institute of Chemical Engineers Journal*, **1986**, *32* (2), 276-290.

Furman, K. C., Sahinidis, N. V. A Critical Review and Annotated Bibliography for Heat Exchanger Network Synthesis in the 20th Century. *Industrial and Engineering Chemistry Research.* **2002**, *41* (10), 2335-2370.

Gundersen, T., Naess, L. The synthesis of cost optimal heat exchanger networks - An industrial review of the state of the art. *Computers and Chemical Engineering.* **1988**, *12* (6), 503-530.

Hohmann, E.C. Optimum networks for heat exchange. Ph.D. Thesis, University of S. California, Los Angeles. **1971**.

Jezowski, J. Heat exchanger network grassroots and retrofit design. The review of the state-of-the-art: Part I. Heat exchanger network targeting and insight based methods of synthesis. *Hungarian Journal of Industrial Chemistry*, **1994a**, *22* (4): 279-294.

Jezowski, J. Heat exchanger network grassroots and retrofit design. The review of the state-of-the-art: Part II. Heat exchanger network synthesis by mathematical methods and approaches for retrofit design. *Hungarian Journal of Industrial Chemistry*, **1994b**, *22* (4): 295-308.

Kern, D.Q. *Process Heat Transfer*. McGraw Hill: New York, **1950**.

Linnhoff, B., Ahmad, S. Supertargeting: Optimum synthesis of energy management systems. *Trans. ASME, J. Energy Resources Technol.*, **1989**, *111*(3), 121-130.

Linnhoff, B., Ahmad, S. Cost optimum heat exchanger networks: I. Minimum energy and capital using simple models for capital costs. *Computers and Chemical Engineering*, **1990**, 14(7), 729-750.

Linnhoff, B., Flower, J.R. Synthesis of heat exchanger networks: I. Systematic generation of energy optimal networks. *American Institute of Chemical Engineers Journal*, **1978**, 24, 633-642.

Linnhoff, B., Hindmarsh, E. The pinch design method for heat exchanger networks. *Chemical Engineering Science*, **1983**, *38* (5), 745-763.

Mizutani, F. T., Pessoa, F. L. P., Queiroz, E. M., Hauan, S., Grossmann, I. E. Mathematical Programming Model for Heat-Exchanger Network Synthesis Including Detailed Heat Exchanger Designs. 2. Network Synthesis. *Industrial and Engineering Chemistry Research,* **2003**, 42, 4019-4027.

Ponce-Ortega, J. M., Serna-Gonzalez, M., Jiménez-Gutiérrez, A. Heat exchanger network synthesis including detailed heat exchanger design using genetic algorithms. *Industrial and Engineering Chemistry Research*, **2007**, *46* (25), 8767-8780

Polley, G.T., Panjeh Shahi, M. H. M. Interfacing heat exchanger network synthesis and detailed heat exchanger design. *Chemical Engineering Research and Design,* **1991**, *69* (3), 445-457.

Shenoy, U.V. Heat exchanger network synthesis: the pinch technology based approach, Gulf Publishing Co., Houston, USA, **1995**.

Townsend, D.W., Linnhoff, B. Surface area targets for heat exchanger networks. Paper presented at *IChemE 11th Annual Res. Meeting*, Bath, U.K. (1984)

Yee, T. F., Grossmann, I. E. Simultaneous Optimization Models for Heat Integration – II Heat Exchanger Network Synthesis, *Computers and Chemical Engineering*. **1990**, *14* (10), 1165-1184.

An MINLP Formulation for the Optimal Design of Distributed Effluent-Cooling Systems

Eusiel Rubio-Castro, Medardo Serna-González, José María Ponce-Ortega

Universidad Michoacana de San Nicolás de Hidalgo, Department of Chemical Engineering, Morelia, México

Abstract

This paper presents a mathematical programming approach to the optimal design of cooling systems for reducing the temperatures of a set of aqueous effluent streams to environmental discharge limit for the effluent temperature. A superstructure is proposed that involves all options of stream bypassing, splitting and mixing along with a cooling tower for effluent thermal treatment. Stream bypasses can reduce both capital and operating costs by lowering effluent flowrate to the cooling tower. The optimization of this structure is formulated as a Mixed Integer Non-Linear Programming (MINLP) problem, which is solved with an implementation of the outer approximation/augmented penalty algorithm. The objective function is to minimize the total annual cost, which accounts for the capital cost for the cooling tower and the operating cost due to power consumption of the tower fan. A detailed model for the cooling tower based on the Merkel's method is used. Therefore, the MINLP model allows simultaneous optimization of the system configuration along with the cooling tower design parameters and operating conditions. Three cases of study are used to show the application of the proposed method.

Keywords: *Aqueous Effluents, Thermal Treatment, Cooling System, Optimization, MINLP.*

Introduction

Chemical, petrochemical and other process industries require significant quantities of freshwater and also discharge much wastewater into the environment. Recently, the increasing scarcity of freshwater resources has motivated process industries to maximize water recycle/reuse to minimize its consumption. In addition, because the increasingly stringent regulations for environment protection, it is also necessary to reduce the chemical and thermal pollution of industrial wastewater to an acceptable level before being discharged into different water disposal sites.

Wastewater streams can be polluted by many soluble organic and inorganic components besides insoluble liquid or solid particles that need to be strictly controlled. Over the past decades, a variety of chemical, physical and biological treatment processes have been developed and practiced in the industries to control the chemical pollution problems of aqueous effluents before disposal (Belhateche, 1995). On the other hand, when the temperatures of wastewater streams from industrial processes are higher than the upper effluent-temperature limit, a thermal treatment system is needed to reduce the temperatures of the aqueous effluents prior to discharge. Based on studies to know the effects of heated-water discharge temperatures on the survival of aquatic organisms, governments set the environmental discharge limit for the effluent temperature which is region specific as well as species specific (Kennedy, 2004).

Several cooling systems can be used to reduce the temperature of the aqueous effluents from industrial sources. Current practice usually merges all the separate effluent streams and utilizes a cooling tower to cool down the single effluent stream generated to the desired temperature before release. Such a system is depicted for three effluent streams in Figure 1. This method for thermal treatment of effluents is referred to as a central cooling system. Kim et al. (2001) proposed a methodology to design distributed cooling systems for effluent temperature reduction. As shown in Figure 2, distributed cooling systems segregate effluent streams for thermal treatment and only combine or bypass them if it is appropriate. The major advantage of distributed cooling systems is segregation of effluents before treatment that leads to processing of small effluent flowrates to the cooling tower with a high driving force for cooling. This could lead to lower total annual cost when compared to central cooling systems. These authors proposed a two-stage approach based on the pinch analysis technology. In the first stage, a mathematical programming model for targeting provides the optimal inlet effluent flowrate to the cooling tower without the need of developing a cooling-system configuration. The objective function is the minimization of the total annual cost (TAC), which is the sum of the cooling tower capital cost and the electricity cost for the operation of water pumping and air fans. In the second stage, the distributed cooling system structure is then obtained manually to reach the optimal target using the grouping design rules. This approach has been a major step in understanding the design of distributed cooling systems. Nevertheless, it relies on a very simplistic short-cut model for the cooling tower and, therefore, it can limit the optimality and practical feasibility of the derived designs. To overcome this limitation, recently Ataei et al. (2009) address design of distributed effluent-cooling systems with a detailed mathematical model of the cooling tower. This method is also based on the pinch analysis technology. Consequently, due to the sequential nature of this approach with targeting followed by design, the trade-off among the investment and operating costs is not considered rigorously.

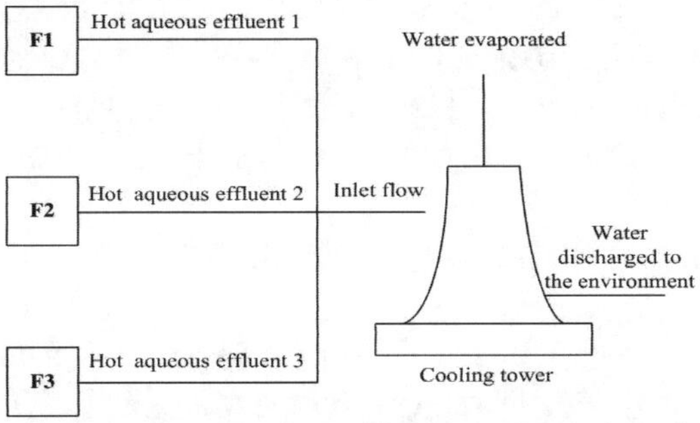

Figure 1. Central cooling system for three aqueous effluents cooling

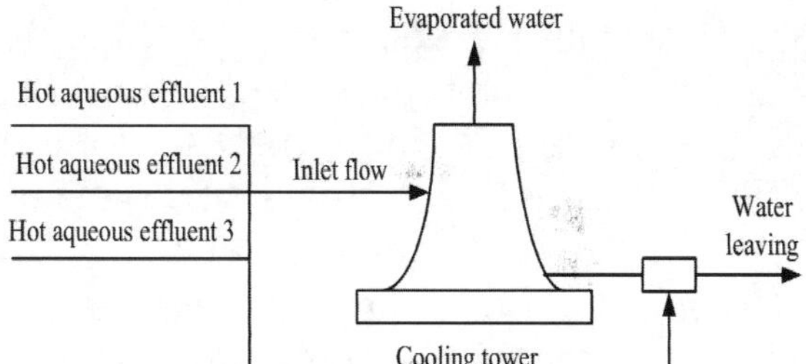

Figure 2. Distributed cooling system for three aqueous effluents

In this work, the problem of temperature reduction of aqueous effluents before discharge is formulated as an optimization model based on a superstructure that embeds all possible cooling system configurations. The superstructure is modeled as a mixed-integer nonlinear programming (MINLP) formulation, which uses a detailed model for the cooling tower. Discrete choices involve deciding what type of packing and type of draft to use for the cooling tower. The objective is to minimize the total annual cost by simultaneous optimization of the cooling system configuration as well as cooling tower design parameters and operating conditions.

2. Model Formulation

The model formulation is based on the superstructure shown in Figure 3. The basic components of the superstructure are stream mixers, stream splitters and a wet-cooling tower.

As an example, the superstructure for the representation of a cooling system for three aqueous effluents that require temperature reduction is presented in Figure 3. Each aqueous effluent has a splitting point followed by two mixing points. Thus, as shown in Figure 3, each aqueous effluent can be split into two streams; one is directed to the mixer at the inlet of the cooling tower while the other is a by-pass stream that is directed to the mixer prior to the final effluent discharged from the plant. It is interesting to note that the superstructure model provides the degrees of freedom available through the by-passing capabilities so as to reduce the effluent flowrate to the cooling tower. In this way, the superstructure features three alternative structural options which are stream mixing, splitting and by-passing. This representation, therefore, contains all the potential cooling system configurations of interest.

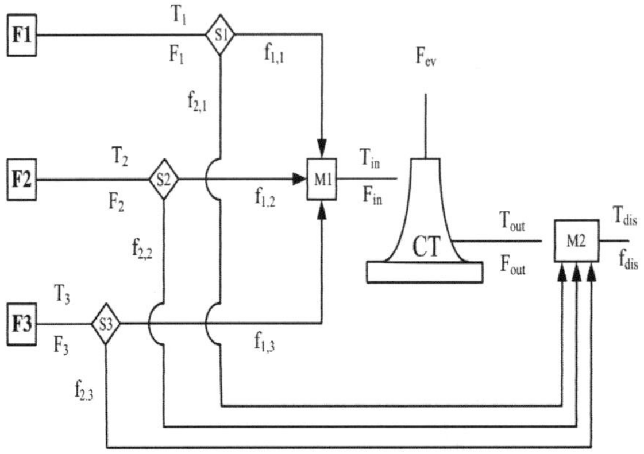

Figure 3. Superstructure for three hot aqueous effluents.

In this section, the superstructure described above is formulated mathematically as an MINLP problem. Because the cooling tower is fixed, the discrete choices consist of deciding what type of packing and type of draft to use for this equipment. The mixers and splitters are also considered as permanent equipment. The selection of flows in the splitters determines the configuration of the optimal cooling system.

To derive the mathematical formulation of the MINLP problem, the effluent streams *j* at the inlet of the superstructure are denoted by the index set $NEF = \{j\}$. Each of these streams *j* has associated as variables the total flowrate Fj and the temperature Tj. The subscripts *out* and *in* refer to the outlet and inlet stream conditions of the cooling tower, respectively. Total flowrate and temperature variables of the final discharge stream have the subscript *dis*. The flowrates of the streams at the outlet of the splitters are labeled with the following conventions:

$f_{1,j}$ = stream from the splitter for effluent stream *j* to the mixer preceding the cooling tower.

$f_{2,j}$ = stream from the splitter for effluent stream *j* to the mixer at the end of the superstructure.

In addition to the variables representing the flowrates and temperatures of the streams, the superstructure contains the following design parameters of the cooling tower: evaporation loss, air mass flowrate, packing height and tower cross-sectional area, type of packing, and power consumption of the tower fan.

Having defined the index set and variables that describe the superstructure, the constraints that apply are as follows:

1. Mass balances at the splitters

$$F_j = f_{1,j} + f_{2,j} \qquad j \in NEF \tag{1}$$

2. Mass balance at mixing point at the inlet of the cooling tower

$$F_{in} = \sum_{j \in NEF} f_{1,j} \quad (2)$$

3. Mass balance at mixing point at the end of the superstructure

$$F_{dis} = F_{out} + \sum_{j \in NEF} f_{2,j} \quad (3)$$

4. Energy balance at mixing point at the inlet of the cooling tower

$$T_{in} F_{in} = \sum_{j \in NEF} T_j f_{1,j} \quad (4)$$

5. Energy balance at mixing point at the end of superstructure

$$T_{dis} F_{dis} = T_{out} F_{out} + \sum_{j \in NEF} T_j f_{2,j} \quad (5)$$

The above equations are restricted by

$$f_{1,j} \leq F_j \quad j \in NEF \quad (6) \qquad F_{in} \leq \sum F_j \quad j \in NEF \quad (8)$$

$$f_{2,j} \leq F_j \quad j \in NEF \quad (7)$$

Cooling tower model

Serna-González et al. (2009) developed an optimization model for mechanical draft counter flow cooling towers that is incorporated in this work as part of the optimization of effluent-cooling systems. The cooling tower model is based on the Merkel's method. Empirical correlations for the air pressure drop and overall mass transfer coefficients in the packing region of the tower are used (Kloppers and Kröger 2003, 2005). Design guidelines and constraints are incorporated to obtain practical cooling towers. The major variables are the air and water mass flowrates, water inlet and outlet temperatures, wet- and dry-bulb temperatures of the air entering the tower, heat load of the tower, type of packing, type of draft, height and area of the tower packing, total pressure drop due to the air flow rate, and power consumption of the fan. Thus, the model includes the selection of type of packing and type of draft using logical decisions through disjunctive programming. Constraints are implemented from the energy balances, heat load equation, Merkel's equation, transfer coefficient and loss coefficient correlations for the tower, relationships of thermodynamic properties and pressure drops along the air path, water consumption equation, practical limits for water mass flowrates, and operational limits and logical constraints.

Water evaporated

The water loss in the cooling tower mainly is yielded by evaporation. The water is also lost from drift and blow-down, but in this study, it is assumed that water losses, except evaporation, are negligible. The water evaporated is calculated by,

$$m_{wev} = m_a \left(w_{out} - w_{in} \right) \quad (9)$$

Objective function

The objective function involves the minimization of the total annualized cost for equipments and utilities,

$$\text{minimize TAC} = k_f CAP + COP \quad (10)$$

where k_f is the annualized factor for investment on cooling tower, CAP is the investment cost for the cooling tower and COP is the operating cost for the cooling tower.

The operating cost involves basically the electricity utility needed for the operation of water pumping and air fan of the cooling tower. Electricity cost needed to operate the air fan is calculated using the following expression:

$$COP = H_Y c_e Pe_f \tag{11}$$

The capital cost for the cooling tower is calculated using the cost function given in Kintner-Meyer and Emery (1995):

$$CAP = C_{CTF} + C_{CTV} A_{fr} L_{fi} + C_{CTMA} m_a \tag{12}$$

This correlation properly reflects the influence of the type of packing (C_{CTV}), the air mass flowrate (m_a) and basic geometric parameters, such as height (L_{fi}) and area (A_{fr}) of the tower packing.

This paper allows the optimal selection of type of packing. This choice is limited to film, splash, and trickle types of fills, which has different costs as well as mass transfer and pressure drop characteristics. Disjunctive programming (Vicchietti et al., 2003) is used to introduce the discrete choices of type of packing both in the objective function and the cooling tower model. The mass transfer and pressure drop characteristics of these types of packing are modeled with the empirical correlations given by Kloppers and Kröger (2003, 2005). Values for C_{CTV} are presented by Serna-González et al. (2009).

3. Results

This section presents the application of the proposed synthesis approach. All the examples were implemented in the software GAMS (Brooke et al., 2006) and solved on a Pentium 4 PC with 2.8 GHz and 512 MB memory. DICOPT was employed for solving the MINLP problems whereas CONOPT was used to solve the NLP subproblems and CPLEX was used for the MILP mater problems.

Example 1. The first example illustrates a case of three effluent streams, which data are given in Table 1. The values for c_e, C_{CTF}, C_{CTMA} and η_f are US$0.085/kWh, US$31,185, US$1,097.5/(kg dry air/s) and 0.75, respectively. The annualized factor and the hours which operate the plant annually are 0.2983 year^{-1} and 8150 hr/year, respectively. The environmental discharge limit for the effluent temperature is taken to be 30ºC.

The effluent-cooling system obtained is shown in Figure 4. Information about the cost data appears in Table 2. This cooling system has a distributed configuration and features a minimum total annual cost by $120,214/year. Note that the optimal inlet flowrate to the cooling tower is generated by mixing effluents 1 and 2. Effluent 3 flows directly to the final mixer. Outlet stream of the cooling tower and effluent 3 are collected in the final mixer to be discharged.

Table 1. Effluent stream data for Example 1

Effluent	T_i (ºC)	F_i (kg/s)
1	60	83.333
2	45	83.333
3	32	111.111

Table 2. Results for Example 1

Configuration	TAC ($/year)	Capital cost	Operation cost	Type of packing	m_a (kg/s)	Power consumption (hp)	Pressure drop (Pa/$m_{packing}$)
Central	136,989.04	106,385.45	30,603.60	Film	181.07	64.74	88.92
Distributed	120,214.58	90,700.12	29,514.46	Film	137.48	64.10	89.40

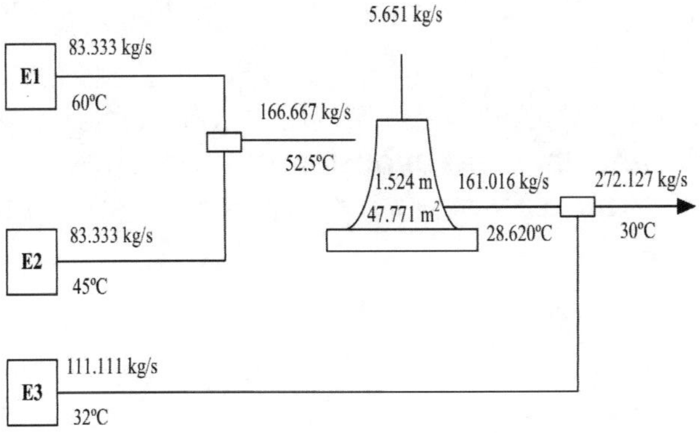

Figure 4. Optimal configuration for Example 1

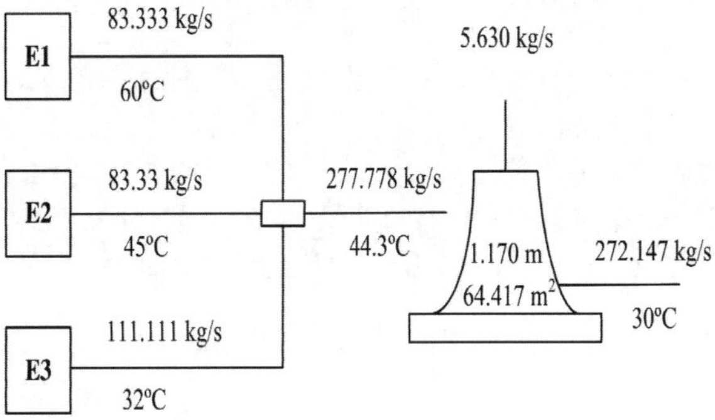

Figure 5. Central cooling system for Example 1

The central configuration is also shown in Figure 5 for this problem. This cooling system features a total annual cost of \$136,989/year, which is 13.95% greater than the optimal solution provided in this paper. The type of packing is the same for both configurations.

Example 2. This example involves deriving a cooling system for five hot effluent streams. The data for streams are given in Table 3. The values for c_e, C_{CTF}, C_{CTMA} and η_f are US\$0.085/kWh, US\$31,185, US\$1,097.5/(kg dry air/s) and 0.75, respectively. The annualized factor and the operation time of the plant are 0.2983 year^{-1} and 8150 hr/year, respectively. Again, the effluent temperatures have to be reduced to 30ºC before the final effluent stream is discharged into the environment.

Table 3. Effluent stream data for Example 2

Effluent	T_j (ºC)	F_j (kg/s)
1	60	27.778
2	55	27.778
3	50	55.556
4	40	41.667
5	35	27.778

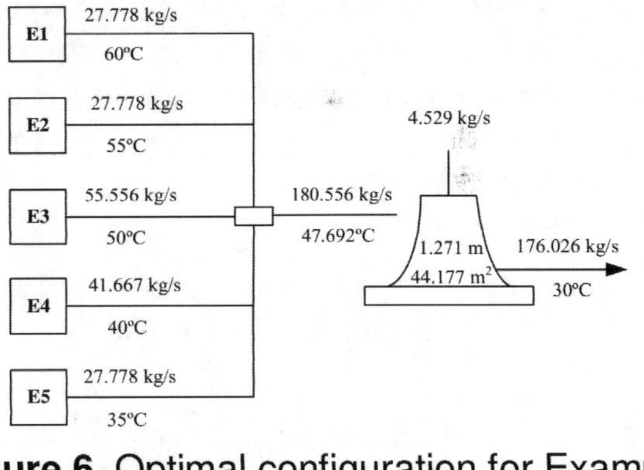

Figure 6. Optimal configuration for Example 2

Table 4. Results for Example 2

Configuration	TAC (\$/year)	Capital cost	Operation cost	Type of packing	m_a (kg/s)	Power consumption (hP)	Pressure drop (Pa/$m_{packing}$)
Central	152,084.31	127,186.03	24,898.28	Film fill	246.519	50.81	76.072

The optimal cooling system is shown in Figure 6. It should be noted that the system has a central configuration. The minimum total annual cost associated with this system is \$152,084/year. Therefore, using the proposed synthesis approach, optimal effluent-cooling systems can be identified that do not have a distributed configuration. This is a very important characteristic since it can lead to economic savings for cases that can be described by capital cost correlations given by Equation (12).

Example 3. Table 3 shows the data for five hot effluent streams by Kim et al. (2001). The values for c_e, C_{CTF}, C_{CTMA} and η_f are US$0.085/kWh, US$31,185, US$1,097.5/(kg dry air/s) and 0.75, respectively and the annualized factor and the operation time of the plant are 0.2983 $year^{-1}$ and 8150 hr/year, respectively. As in examples 1 and 2, the environmental discharge limit for the effluent temperature is 30ºC.

Table 5. Effluent stream data for Example 3

Effluent	T_i (ºC)	F_i (kg/s)
1	40	55.556
2	35.3	196.667
3	35	486.111
4	27	64.444
5	25	111.111

The results of the optimization problem provide the distributed cooling system shown in Figure 8. It can be seen that only the splitter at the third effluent stream is present in the optimal system structure, while the other splitters are discarded by the optimization procedure. That splitter splits effluent 3 into one stream that is directed to the mixer at the inlet of the cooling tower and into one bypass stream. Effluents 1 and 2 go directly to the mixer preceding the cooling tower while effluents 4 and 5 flow completely to the final mixer.

As shown in Table 6, the optimal distributed cooling system features a minimum TAC by $152,084/year, which is significantly lower than the cost of $183,564/year of the best configuration that was obtained by Kim et al. (2001) and it is presented in Figure 7. In addition, this solution is derived via an optimization approach in which the flowrates and temperatures of the different streams are determined simultaneously together with the details of the cooling tower. The cost of the distributed design obtained in this paper ($152,084/year) is also substantially lower than the cost of the central cooling system shown in Figure 9 ($202,454/year).

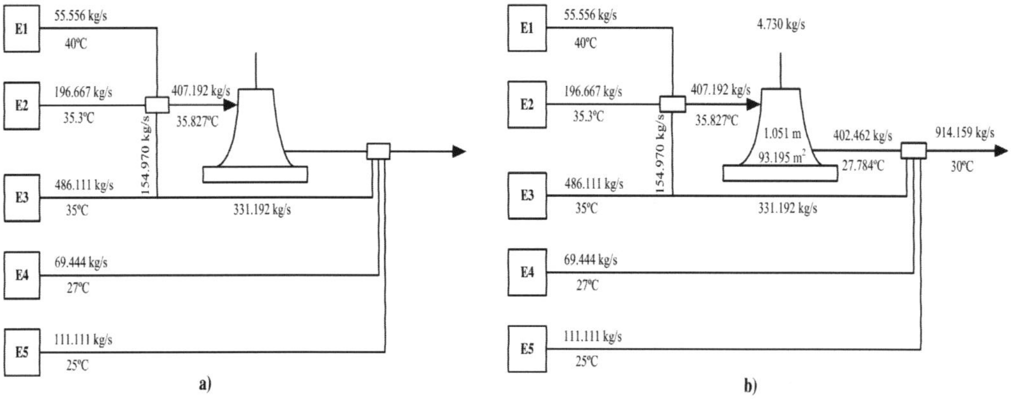

Figure 7. Best configurations for Example 3 from Kim et al. (2001)

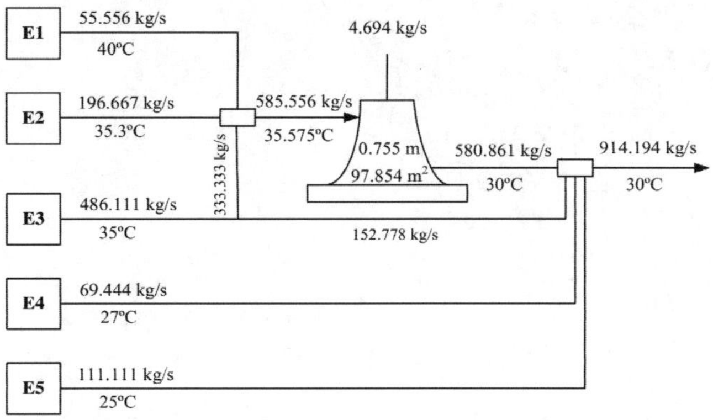

Figure 8. Optimal structure for Example 3 obtained in this paper

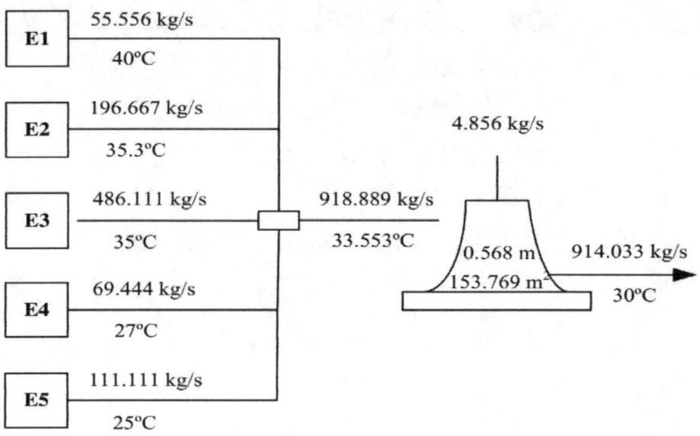

Figure 9. Central structure for Example 3

Table 6. Results for Example 3

Configuration	TAC ($/year)	Capital cost	Operational cost	Type of packing	m_a (kg/s)	Power consumption (hP)	Pressure Drop ($Pa/m_{packing}$)
Central	202,453.96	178,888.31	23,565.65	Film	366.584	47.146	56.151
Kim et al. (2001)	183,564.06	143,874.73	39,689.33	Film	261.96	80.74	86.49
This paper	152,084.31	127,186.03	24,898.29	Film	246.52	50.81	76.072

4. Conclusions

This paper presents an MINLP model based on a superstructure that allows splitting and bypassing of aqueous effluents to find the optimal cooling system for effluent temperature reduction. A particular feature of this approach is that it uses a detailed model for the cooling tower during the optimization, which is based on the Merkel's method. The most important variables of the cooling tower are: cross-sectional area for packing section, tower fill height, type of packing, pressure drop through the fill, and water and air flow rate.

The objective function is to minimize the total annual cost, which involves the capital cost for the cooling tower and the operating cost due to power consumption of the tower fan. Solution of this mathematical formulation provides simultaneously the optimal cooling system configuration, which satisfies the specified environmental discharge limit for effluent temperature, as well as the optimal operating conditions and design parameters required for the cooling tower. Therefore, the proposed approach overcomes the limitations of previously reported methodologies based on pinch technology.

Three example problems are presented to show the application of the proposed MINLP model. Results indicate that considering simultaneously the trade-offs between the design variables is a very important task to yield the configuration with the minimum TAC.

Nomenclature

c unity cost, US$/kgWatts, US$/Joules
C_{CTF} initial cost of the cooling tower, US$
C_{CTMA} cost related to flow, US$ s/kg
C_{CTV} cost of the cooling tower due to packing volume, US$/m^3
COP operation cost, US$/year
CAP capital cost, US$/year
H_Y hours that operates the plant annually, hr/year
k_f annualized factor, yr^{-1}
f flowrate of connecting streams, kg/s

F	flow rate of effluent streams, kg/s
m	flow rate, kg/s
NEF	number of aqueous effluents
Pe	power consumption, W
T	temperature, ºC
TAC	total annual cost, US$/year
w	mass fraction humidity, kgwater/kgdryair

greek symbols

η_f	fan efficiency

index

a	air, dry-bulb
dis	discharge
e	electricity
ev	evaporate
f	fan
in	inlet
out	outlet
w	water

References

Ataei, A., Panjeshahi, M.H. Gharaie, M. and Tahouni, N., 2009, New method for designing optimum distributed cooling system for effluent thermal treatment. *International Journal of Environmental Research*, 3(2), 155-166.

Belhateche, D. H. 1995, Choose appropriate wastewater treatment technologies. *Chemical Engineering Progress*, August, 32.

Brooke, A., Kendrick, D., and Meeraus, A., 2006, GAMS User's Guide, The Scientific Press, USA.

Kennedy, V. S. 2004, Thermal pollution. Encyclopedia of Energy, Volume 6. Elsevier Inc.

Kim, J., Savulescu, L. and Smith, R., 2001, Design of cooling systems for effluent temperature reduction, *Chemical Engineering Science*, 56, 1811-1830.

Kintner-Meyer, M., and Emery, A.F., 1995, Cost-optimal design for cooling towers, *ASHRAE J*, **4**, 46-55.

Kloppers, J.C., and Kröger, D.G., 2003, Loss coefficient correlation for wet-cooling tower fills. *Applied Thermal Engineering,* **23**(17), 2201-2211.

Kloppers, J.C., and Kröger, D.G., 2005, Refinement of the transfer characteristic correlation of wet-cooling tower fills, *Heat Transfer Engineering,* **26**(4), 35-41.

Vicchietti, A., Lee, S., and Grossmann, I.E., 2003, Modeling of discrete/continuous optimization problems characterization and formulation of disjunctions and their relaxations. *Computers and Chemical Engineering*, **27**(3), 433-448.

Serna-González, M., Ponce-Ortega, J.M., and Jiménez-Gutiérrez, 2009, A. MINLP optimization of mechanical draft counter current wet-cooling towers. *Chemical Engineering Research and Design*, submitted for publication.

A Process Synthesis Approach to Improving the Efficiency of Coal Fired Power Stations

Baraka Celestin Sempuga, School of Chemical and Metallurgical Engineering, Centre of Material and Process Synthesis, University of the Witwatersrand, Johannesburg, South Africa

Diane Hildebrandt, School of Chemical and Metallurgical Engineering, Centre of Material and Process Synthesis (COMPS), University of the Witwatersrand, Johannesburg, South Africa

David Glasser, School of Chemical and Metallurgical Engineering, Centre of Material and Process Synthesis (COMPS), Johannesburg, South Africa

Bilal Patel, Comps, Wits University, Johannesburg, South Africa

A large portion of the energy used worldwide originates from fossil fuels. Coal in particular is the major resource for mechanical and electrical power production. The use of coal for power production involves direct combustion of coal for high pressure and high temperature steam generation followed by mechanical work recovery by passing the steam through turbines.

However this process has been proven to be highly inefficient, leading to the loss of the large portion of the coal work potential.

In this work a graphical approach is used to show that most of the inefficiencies lie in the stage of coal combustion. It can be shown that considerable improvement can be made by changing the chemical route for coal combustion; this could result in the production of valuable chemicals in addition to the work produced.

One of the chemical routes explored constitutes partial oxidation of coal combined with the catalytic production of hydrogen, followed by ammonia production. The ammonia can then be oxidised catalytically to nitric oxides for subsequent production of nitric acid and fertiliser. It is shown that each step in this process has a certain amount of work potential which can be recovered through steam production. It is also shown that in addition to the nitric acid production, approximately 10% improvement in work production efficiency can be achieved; compared to the direct coal combustion where only carbon dioxide is produced.

By combining fertiliser manufacture with electricity production one can in principle improve the efficiency of both processes. This is potentially a real win-win situation.

See more of this Session: Innovations in Energy Processes
See more of this Group/Topical: Process Development Division

Energy Efficiency Improvement in LNG Liquefaction Processes

Minsu Kim, Chemical and Biomolecular Engineering, Yonsei University, Seoul, South Korea

Wonsub Lim, Chemical and Biomolecular Engineering, Yonsei University, Seoul, South Korea

Daeho Ko, GS Engineering & Construction, Seoul, South Korea

Kwangho Choi, GS Engineering & Construction, Seoul, South Korea

Il Moon, Chemical and Biomolecular Engineering, Yonsei University, Seoul, South Korea

This paper focuses on the efficiency comparison of various LNG liquefaction processes and suggests a better process which requires less power and handles larger MTPA using simulation and optimization techniques. Our team analyzed steady-state processes such as Single-Mixed Refrigerant processes and Cascade Single-Mixed Refrigerant processes case by case. Recent studies forecast the market demand of the global LNG will increase approximately 10% in next 10 years. And the global LNG plant market will increase as LNG consumption increases.

Major simulation variables in SMR refrigerant systems are the selection of equipment and compositions of feed gases. Another important variable is the compressor ratio changed for decreasing required power at the same capacity. Many other variables are considered to decrease the power consumption. The cascade SMR processes are also analyzed. The capacity of LNG production increases with low power consumption. Peng-Robinson like equations are used to solve thermodynamic properties. To improve its accuracy, our team measured various properties under some critical conditions. These experimental data are imbedded in commercial physical property routines and they improve the accuracy of the simulation. Taken some operating parameters as key parameters, influences of these compressor ratios on shaft power and compositions of mixed refrigerant are analyzed. A new efficient LNG liquefaction process is developed in this research.

See more of this Session: Innovations in Energy Processes
See more of this Group/Topical: Process Development Division

Conversion of Alaskan Natural Gas to a Liquid Fuel Using Microchannel Reactor/Heat-Exchangers

Stuart W. Churchill[1], Jeff Hammond[2], Jared Lee[2], Moyd. Shayaan Nadeem[2] and Sophie Weiss[2], (1)Chemical & Biomolecular Engineering, The University of Pennsylvania, Philadelphia, PA, (2)The University of Pennsylvania, Philadelphia, PA

Conversion of natural gas to a mixture of liquid hydrocarbons (usually in the form of a synthetic light crude oil) by means of chemical reaction is known by the acronym GTL. A new GTL process is proposed with the objective of bringing to the market the energy in the proven reserve of 12E12 cubic feet in a "stranded" field on the North Slope of Alaska near Prudhoe Bay. The liquid would be pumped through the nearby existing 800-mile-long Trans Alaskan Pipeline System (TAPS) to Valdez on Prince William Sound where it would be stored and then transported to a refinery by tanker.

A number of gas-to-liquid (GTL) processes have been proposed and a few have actually been utilized. What is unique in the proposed process is the use of modular microchannel reactor/heat-exchangers to carry out two sets of solid-catalyzed reactions: steam-methane reforming (which is highly endothermic) at 255 $\rightarrow$ 754 oF and ~225psig; and the Fischer-Tropsch conversion (which is highly exothermic) at 437 oF and ~480 psig.

The isolated location, the extreme environmental conditions, and the absence of the usual supporting services also influenced the design of the modified process. Prudhoe Bay, which is within the Arctic Circle, is open to shipping only six months a year. The site of the proposed plant has an extreme climate including periods of continuous night and continuous daylight. Because of the absence of any nearby habitation or industry there is not an existing supply of electricity, non-saline water, and other such materials and services.

The design was based on the conversion of 7.63E9 CFD of natural gas to 118,000 barrels of liquid hydrocarbons for the pipeline. The choice of the modular (rectangular, stacked) microchannel reactors developed by Velocys was based on the prospect of improved rates of heat transfer and reaction, and thereby a reduction in the size of the equipment and in the cost of shipping it to such a remote site..

The required raw materials are 7.74E9 CFD of natural gas, including 4.38E8for combustion in the reactor and 9.9E6 for the generation of electricity, as well as 7.63E6 lb/hr of water for reaction, startup, and makeup for losses. It was postulated that this water would need to be purchased from a distant site. The outputs other than liquid hydrocarbons include some CO2 to be sequestrated, used in oil recovery or discharged to the atmosphere, and some waste-water with a trace of dissolved hydrocarbons.

For the reforming step, a mixture of steam and methane was passed through a set of passages in a module where it was preheated by product-to-feed exchange, further heated by exchange with hot burned gas in passage, and finally reacted. The products of reaction

(Syngas) were cooled in countercurrent flow through intermediate passages of the same module. A mixture of methane and air passed through a third set of intermediate passages where it was preheated by exchange with the burned gas and after burning was cooled in passage through a fourth set of countercurrent passage by exchange with the reacting mixture and then with the entering methane and air. The modules were 1.5m high, 1m wide, and 0.6m long. The passages were 0.25 mm in height and the spacers, including the layer of catalyst were mm in height. Thermo-kinetic calculations indicated that 3298 modules operating in parallel were required. Similar modules were utilized for the Fischer-Tropsch conversion of the Syngas. In this case the reaction took place in 0.95mm passages and water was boiled in alternating 1.5mm passages. Thermal and chemical-kinetic modeling indicated that 3987 modules operating in parallel were required.

The separations can be summarized briefly as follows. CO2 is removed from the Syngas using absorption in MEA. A first flash and decanting separates gaseous, liquid hydrocarbons, and wastewater, and a second flash removes most of the residual gases. CO is separated from the gas from the first flash by the COPureTR process. The CO and the residual gas are recycled to the reformer.

The electrical requirement for compression, pumping, etc. is 782MW so a 800MW generator using natural gas was considered an integral part of the process.

Fixed and variable costs were based on a site factor of 1.4, a 25-year operation, a raw natural gas cost of $0.80/MSCF, and a tariff for TAPS of $4.87 per barrel.The total capital cost was estimated to be $7.17B, the annual fixed cost to be $572MM, the annual variable cost to be $106MM. At a discount rate of 13% an IRR of 15% and an NVP of $708MM. These cost estimates are very sensitive to the price of crude oil, which is difficult to estimate next year let alone in 25 years.

How do these predictions differ in technical and economic promise from those for (1) direct liquefaction and shipping of natural gas, (2) other gas-to-liquid reactive schemes, and (3) pumping the natural gas to the market through a pipeline?

Direct liquefaction does not appear to be feasible at this site because Prudhoe Bay is open to shipping only six months of the year. A large storage capacity would be required. The facilities for direct liquefaction would be subject to many of the same difficulties associated with isolation and climate.

Construction of a 1300-mile pipeline from Prudhoe Bay to connect in Alberta with the Canadian natural gas network has been proposed. It would have a capacity of 4.5E9 cubic feet per day and an estimated capital cost of $26B, which is far greater than that for the GTL plant on the basis of equal quantities of natural gas per day. The initial pressurizing and the intermediate repressurizing would require dedicated electrical generators driven by combustion of some of the natural gas from the pipeline.

.

Reactive conversion of natural gas to a liquid has a long history, including the following ventures. Sasol (Chevron) has been operating plants using a slurry-reactor, one with an output of 124,000 BPD, in South Africa since 1970. ORYX GTK. A joint venture of Qatar Petroleum and Sasol is said to be planning to increase the capacity of their low-temperature Fischer-Tropsch GTL plant in Qatar to 100,000 BPD. Shell Oil is planning to open a 140,000 BPD GTL plant called "Pearl" in Qatar in 2010. In 2003 the capital cost was estimated to be $4-6B but in 2007 that was increased to $12-18B. In February 2007, ExxonMobil and Qatar Petroleum cancelled plans for a 154,000 BRD GTL plant with an estimated cost of $7B. BP operates a demonstration GTL plant in Nikishi, near Anchorage, Alaska, that produces 300BPD. It utilizes compact rather than microchannel reactors.

In conclusion, the new GTL process appears to be technically advantageous over prior ones for the North Slope site. In particular, the reductions in volume and weight of 84% and 78%, respectively, for the reactors, the reduction of the exit temperature of the reformer from 1626 to 756 oF, and the increased catalytic productivity (lbprod/lbcat/hr) of the Fischer-Tropsch conversion from 0.07 to 1.53. Whether or not these gains are sufficient to favor the process over the construction of a natural gas pipeline is uncertain because of the uncertainty of the price of oil.

Abstract ID#: 157688
Password: 343188
Title: Conversion of Alaskan Natural Gas to a Liquid Fuel Using a Microchannel Reactor/Heat-Exchangers
Session Selection: 12G00 Design of Alternative Fuel Products
Preferred Presentation Format: Oral
Lead Presenter's Email: churchil@seas.upenn.edu

Related to existing technology or advanced technology associated with the overall use and supply of energy or global climate change research? Yes
Is your paper related to fossil, renewable or nuclear energy? Fossil
Specific Type of Technology used: Natural Gas
Research is funded by: Private Non/Not For Profit Organization

First author

Presenting Author
Stuart W. Churchill
Professor Emeritus
Chemical & Biomolecular Engineering
The University of Pennsylvania
220 South 33rd St
311A Towne Bldg.

Philadelphia, PA 19104
Phone Number: 215-898-5579
Fax Number: 215-573-2093
Email: churchil@seas.upenn.edu

Second author

Jeff Hammond
The University of Pennsylvania
220 South 33rd St
311A Towne Bldg.
Philadelphia, PA 19104
Fax Number: 215-573-2093
Email: thammond@seas.upenn.edu

Third author

Jared Lee
The University of Pennsylvania
220 South 33rd St
311A Towne Bldg.
Philadelphia, PA 19104
Fax Number: 215-573-2093
Email: JaredJL@seas.upenn.edu

Fourth author

Moyd. Shayaan Nadeem
The University of Pennsylvania
220 South 33rd St
311A Towne Bldg.
Philadelphia, PA 19104
Fax Number: 215-573-2093
Email: Shayaan@seas.upenn.edu

Fifth author

Sophie Weiss
The University of Pennsylvania
220 South 33rd St
311A Towne Bldg.
Philadelphia, PA 19104
Fax Number: 215-573-2093
Email: Sophiejw@seas.upenn.edu

MULTI-SCALE MODELLING OF REFINERY PRE-HEAT TRAINS: EFFECT OF CRUDE FOULING ON ENERGY RECOVERY, CARBON EMISSION AND ECONOMICS

Francesco Coletti and Sandro Macchietto
Department of Chemical Engineering and Chemical Technology
Imperial College London, South Kensington Campus
London SW7 2AZ, UK

Extended abstract

Pre-heat trains (PHTs) in crude distillation units are complex networks of heat exchangers used to recovery up to 60-70% of all the energy necessary for primary fractionation.[1] Efficiency of the train is severely impaired by fouling, the deposition of unwanted material on the thermal surfaces. Economic losses related to this problem was estimated of the order of 1.2 billion dollars p.a. in the United States alone when the price of oil was significantly cheaper.[2] Maintenance issues as well as health and safety hazards are also related to this problem that oil refineries have been facing for decades.

Experimental evidences show that fouling rates are function of crude composition and process conditions, namely temperature and velocity. It has also been demonstrated that, for a given crude composition, threshold conditions of temperatures and velocities exist below which no fouling occurs.[3] However, the complex chemical and physical mechanisms underlying fouling deposition are still poorly understood. For this reason, both design and monitoring techniques, based on highly empirical fouling factors, have substantially not changed over the past 40 years.

To tackle this long standing problem in a systematic way, a large scale interdisciplinary research project, CROF (for CRude Oil Fouling) has been established.[4] The project, coordinated in 8 sub-projects, aims at providing a platform to investigate fundamental parameters leading to deposition, to provide a framework for predicting deposition and avoiding it by design and to formulate methods for mitigation.

Within the CROF framework, a multi-scale, first principles model for a shell and tube heat exchanger was developed and validated against data from an ExxonMobil refinery.[5] The model is dynamic, distributed and accounts for local variation of physical properties. Several scales of investigation are considered simultaneously. At the tube level, the interactions between operating conditions and fouling are captured by coupling the solution of a moving boundary problem with the use of the Ebert-Panchal[6] model in a distributed way. This distinctive feature of the model, allows calculating the thickness of the fouling layer along the tubes in each exchanger and its interactions with heat transfer and fluid-dynamics. At unit level, it is possible to identify critical zones where deposition is particularly severe. Complex interactions affecting the overall performance of the pre-heat train can then be unveiled by interconnecting several units in a network.

In this work, simulations for the hot end of a refinery pre-heat train are presented showing the effects of fouling on the thermal-hydraulic performance of the network. The rigorous computation of several key indicators such as the reduction in throughput, energy losses, uncontrolled flowsplits, and carbon emissions allowed assessing the impact of fouling on the economics of the refinery. For the future, it is envisaged the use of this approach to optimise design and operations of the network, its cleaning schedules and energy efficiency strategies.

References

1. Panchal, C.B. and Huangfu, E.-P. Effects of Mitigating Fouling on the Energy Efficiency of Crude-Oil Distillation. *Heat transfer engineering,* 2000; 21, (3): 3-9.
2. ESDU *Heat exchanger fouling in the pre-heat train of a crude oil distillation unit, Data Item 00016*; ESDU International plc.: London, 2000.
3. Knudsen, J.G., Dahcheng, L. and Ebert, W.A. *The determination of the threshold fouling curve for a crude oil*, In: Bott, T. R., Ed. Understanding Heat Exchanger Fouling and its Mitigation, Begell House: Lucca, Italy, May, 1997; pp 265–272.
4. Macchietto, S., Hewitt, G.F., Coletti, F., Crittenden, B.D., Dugwell, D.R., Galindo, A., Jackson, G., Kandiyoti, R., Kazarian, S.G., Luckham, P.F., Matar, O.K., Millan-Agorio, M., Müller, E.A., Paterson, W., Pugh, S.J., S.M., R. and Wilson, D.I. *Fouling in Crude Oil Preheat Trains: a Systematic Solution to an Old Problem*, In: Müller-Steinhagen, H.; Malayeri, R.; Watkinson, P., Eds. EUROTHERM conference on Fouling and Cleaning in Heat Exchangers, Schladming, Austria, June 14-19, 2009.
5. Coletti, F. and Macchietto, S. A dynamic, distributed model of shell and tube heat exchangers undergoing crude oil fouling. *Submitted for publication,* 2009.
6. Panchal, C.B., Kuru, W.C., Liao, C.F., Ebert, W.A. and Palen, J.W. *Threshold conditions for crude oil fouling*, In: Bott, T. R., Ed. Understanding Heat Exchanger Fouling and Its Mitigation, Begell House: Lucca, Italy, May, 1997; pp 273-281.

Biogasification of Sugarbeet Tailings: Design, Construction and Operation of a Demonstration Plant

Pratap Pullammanappallil, Doug Renk, Paul Lane, Abhay Koppar and David Chynoweth

Department of Agricultural and Biological Engineering
University of Florida
Gainesville, Fl 32611-0570

Nearly forty percent of all refined sugar consumed in the USA is made from sugar beets grown in the north central and north western regions of the United States. Beet sugar processing generates significant quantities of solid organic waste called tailings. Raw sugar beets are first washed and separated from "tailings" which mainly consist of sugar beets, weeds, sugar beet tops, debris and soils held by sugar beets when harvested. Usually, these tailings are stockpiled outside the factory and hauled away for disposal into landfills or applied on nearby farmland at a significant cost to the factory. American Crystal Sugar Company operates five processing plants along the Red River Valley in Minnesota and North Dakota. It spends close to $1 million per year disposing 400 tons of tailings that are generated daily just at its East Grand Forks plant.

Sugar production from sugar-beets is an energy intensive operation due to the need for drying and evaporation. The East Grand Forks plant for example spends about $1 million per year for natural gas. Anaerobic digestion (or biogasification) of tailings would generate biogas fuel that could potentially reduce the fossil fuel requirements and moreover fuel would be produced on site guaranteeing a fuel supply that is not subject to price fluctuations.

In this paper we will present the development of a biogasification system to anaerobically digest tailings. A demonstration scale high solids anaerobic digestion system to biogasify 2 tons per day of tailings was constructed and is being operated at the American Crystal Sugar Company, Moorhead. The design was based on laboratory scale and pilot scale studies at the University of Florida. The design, implementation, construction and start-up of the system will be presented. In addition, the process performance in terms of organic loading rate, methane production rate, methane yield, methane composition in gas phase, effluent soluble COD, residual volatile organic acids and pH will be presented.

Rheology and Stability Considerations in Processing of Bio-Oils

Xiaoming Pan, Chemical Engineering, University of Massachusetts, Amherst, Amherst, MA

Tushar Vispute, University of Massachusetts - Amherst, Amherst, MA

George W. Huber, Chemical Engineering, University of Massachusetts, Amherst, MA

Surita R. Bhatia, Chemical Engineering, University of Massachusetts, Amherst, Amherst, MA

Understanding the rheology and stability of biofuels has been identified as a major need in the large-scale production and use of these materials. A particularly challenging problem has been quantifying and eliminating the dramatic viscosity increase that can occur upon storage of biofuels. Here, we report results on the viscosity, microstructure, and chemical composition of bio-oils prepared by a fast pyrolysis approach, upon aging these fuels at high temperatures for periods of several days. Our preliminary results suggest that the viscosity increase is not correlated with any microstructural change. In addition, while there have been significant efforts on removing chars as a means of stabilizing biofuels, our results show that removal of neither large nor small char particles has any significant impact on the rate of viscosity increase. The viscosity increase appears to be due to formation of high molecular weight polymeric species over time. Even small amounts of these species (~1 wt%) have a dramatic effect on the rheology. Our work also suggests that hydrogenation of the samples may be beneficial in eliminating the viscosity increase. Overall, these studies help identify the major issues that must be addressed in order for bio-oils to be produced, stored, and transported efficiently in large volumes.

See more of this Session: Innovations in Energy Processes
See more of this Group/Topical: Process Development Division

Sustainable Energy Production: The Great 21st Century Challenge in Chemical Engineering

Eric W. McFarland, Chemical Engineering, University of California, Santa Barbara, Santa Barbara, CA

Massive scale integrated chemical transformations will always be central to producing fuels and chemicals to sustain a prosperous society and this is what chemical engineers do. Whether or not the earth's climate is being adversely affected by humankind's reliance on fossil fuels - they are a finite and their use unsustainable economically, if not environmentally much beyond the present generation. Predicting precisely how and when new energy and chemical production technologies will evolve is an amusing exercise where most everyone will be partially correct and partially wrong. There is little doubt that a combination of nuclear, wind, and solar (biomass, thermal, photovoltaics) will play an increasing role in providing large scale energy supplies for fuels and chemicals; however, the unpredictable mixture of short and long term economic considerations with politics, social psychology, and risk perception will determine society's and industry's response to the challenge of fueling our future. History has shown that bulk energy costs much higher than $10/GJ are associated with disastrous economic consequences. Innovation is predictably unpredictable and bringing to bear the world's technical talent on developing improved conversion technology will undoubtedly lead to unexpected advances which we can hope, but not count on, to bring down costs and maximize the utility of all potential energy sources. Today, the only logical approach is to move forward with multiple options in parallel and adjust our strategy as new innovation and social conditions change. Chemical engineering is the only professional discipline with educational background in the fundamentals of most all the critical elements of any future energy conversion and storage technology and it is the responsibility of chemical engineers to assume leadership roles in developing and executing strategies to evolve our fuel and chemical industries to sustainability so as to minimize the societal disruption and to insure that future generations have the opportunities for greater prosperity than the generations before them. It is also our responsibility to provide society with accurate information and to correct misinformation regarding the benefits and risks of energy conversion technologies. Historical aspects of energy supplies and economics will be coupled to present technologies and specific challenges and opportunity for chemical engineers will be highlighted.

See more of this Session: Innovations in Energy Processes
See more of this Group/Topical: Process Development Division

Photocatalytic Concrete Slab for Passive NOx /O3 Pollution Control

M. Palla, Chemical Engineering, Lamar University, Beaumont, TX

Archana Vaddi, CBI Engineering, Beaumont, TX

R. Asapu, Chemical Engineering, Lamar University, Beaumont, TX

Daniel H. Chen, Chemical Engineering, Lamar University, Beaumont, TX

Robert L. Yuan, Civil Engineering, Lamar University, Beaumont, TX

NOx emissions have become a constraint for sustainable economic development due to ozone non-attainment. Under light illumination over a photocatalyst free radicals are generated to oxidize or decompose pollutants. While volatile organic compounds (VOCs) are oxidized to water, CO2, and chloride, NOx are oxidized to nitric acid, neutralized by the alkaline materials in concrete, and washed away by rain. Ozone can be decomposed to oxygen over photocatalysts. Recent developments in Europe Japan, Hong Kong, and US have demonstrated the interest in deploying photocatalytic technologies for environmental remediation. The photocatalytic coating has potential to economically reduce nitrogen oxides (NOx) emissions at

ambient conditions nearly maintenance-free.

We report here the application of photocatalytic coatings on a concrete slab to simulate road pavements for passive NOx pollution control. Catalysts tested include Degussa P-25 TiO2 and TiO2 - boosted with BaTiO3, Nd2O3, and Fe2O3. Both NO and NO2 oxidation including the effects of inlet NO concentration, residence time, and humidity are investigated. The removal efficiencies are in the range of 86-93% from 200 ppb-2000 ppb NO. Relative Humidity (RH) is seen to have some beneficial effect at <35% RH while lowering NO conversion by ~10% at > 35% RH. The additives Nd2O3, Fe2O3, and BaTiO3 are observed to depress the NO2 generation while improving NO conversion when using fiber-optic as a support. When tested on catalytic concrete slabs, however, Degussa P-25 offers the best NO removal, even though Fe2O3 and Nd2O3 offer better NO2 suppression. The optimal TiO2 to cement ratio is of interest because TiO2:SiO2 photocatalyst shows twin peaks in their activity in atrazine oxidation and the fact that cement also contains Fe2O3. A CFD simulation of NOx plumes from a tail pipe shows the likely distribution of concentration and residence time due to dispersion; that, in turn, can provide a useful guide for future lab investigations.

Catalyst-coated concrete photoreactors with small dimensions of 8" X 4" X 2"were fabricated according to the ACI mix design with a 3-mm catalytic layer with Type 1 Portland cement: Degussa P-25 TiO2 = 2:1 (weight ratio) and TiO2 : SiO2 : CaO : Fe2O3 = 1 : 0.58 : 1.77 : 0.09 (mole % ratio). Two 4 watt UVA (Black Light) gives a total radiation exposure is 380 watt/m2. Catalyst loading is 13 mg/m2 in 3 mm(~1/8") uniform layer. Experiments are conducted to find out the NO oxidation different inlet NO concentrations (200 ppb, 500 ppb, 1500 ppb & 2000 ppb) and different RT (5sec, 10 sec, 15 sec & 20 sec). Ozone is generated using Ozone generating UV lamps and oxygen. Ozone oxidation on the catalyst coated photoreactor boosted with MnO2 at different inlet Ozone concentration and different Residence Time will be tested. Ozone is analyzed using the ozone analyzer (model: 49C O3 analyzer, Thermoscientific Instruments Inc.). HONO measurement to find out whether there is any formation during photocatalysis reaction for both NO and NO2 oxidation leaving the photocatalytic reactor using the HPLC analysis.

See more of this Session: Process and Product Development for Sustainability
See more of this Group/Topical: Process Development Division

Process Design of Wastewater Treatment Plants for Textile Effluents

Ka Y. Fung, Department of Chemical and Biomolecular Engineering, The Hong Kong University of Science and Technology, Hong Kong, China

Chin M. Lee, Department of Chemical and Biomolecular Engineering, The Hong Kong University of Science and Technology, Hong Kong, China

Christianto Wibowo, ClearWaterBay Technology, Pomona, CA

Ka M. Ng, Department of Chemical and Biomolecular Engineering, The Hong Kong University of Science and Technology, Hong Kong, China

Textile industry consumes a humongous amount of fresh water, depleting a valuable resource required for everyday human activities. In addition, wastewater effluents loaded with pollutants often pose significant risks to the environment. It is therefore important to have high-performance and low-cost wastewater treatment plants to treat and reuse textile wastewater. Despite its importance, many wastewater treatment plants are still designed based on experience, leading to high consumption of chemicals, plant failure in the presence of disturbances and inappropriate selection of wastewater treatment techniques such as coagulation, adsorption, biotreatment, Fenton reaction, ozonation, multi-media filtration, membrane filtration, etc.

An integrated design method has been formulated for the systematic development of a wastewater treatment plant. Textile effluents are first characterized to provide basic information for process design. Heuristics are used to guide the designer to come up with a preliminary design. Design and operating parameters of the wastewater treatment plant are then determined based on bench-scale experimental testing. Finally, various flowsheet alternatives are evaluated to select the optimum process flowsheet using a simulation code. The workflow among various stakeholders to reach the final design is also discussed. Industrial examples are presented to illustrate the design method.

See more of this Session: Process and Product Development for Sustainability
See more of this Group/Topical: Process Development Division

Study On Emission Reduction for Chemical Industry Sustainability

Qiang Xu, Dan F. Smith Department of Chemical Engineering, Lamar University, Beaumont, TX

Jian Zhang, Dan F. Smith Department of Chemical Engineering, Lamar University, Beaumont, TX

Xiongtao Yang, Dan F. Smith Department of Chemical Engineering, Lamar University, Beaumont, TX

Chaowei Liu, Dan F. Smith Department of Chemical Engineering, Lamar University, Beaumont, TX

Yanqin Wen, Dan F. Smith Department of Chemical Engineering, Lamer University, Beaumont, TX

Chuanyu Zhao, Dan F. Smith Department of Chemical Engineering, Lamer University, Beaumont, TX

Jie Fu, Dan F. Smith Department of Chemical Engineering, Lamer University, Beaumont, TX

Kuyen Li, Dan F. Smith Department of Chemical Engineering, Lamar University, Beaumont, TX

Emission reduction is one of the major issues for chemical industry nowadays. This is not only because of the increasing stringent environmental regulations and public concerns for air emission, but also due to the fact that emission usually means the loss of process raw material and energy which supposedly could generate much needed products from the industry. Thus, emission reductions have significant impacts on environmental, societal, and economic advancement for chemical industry sustainability.

This paper will summarize the general idea of some best practice technologies for chemical plant emission reductions. The focus of the idea is from the system point of view to effectively use information technology (simulation, optimization, and scheduling) to help plant reduce air emissions. From the system point of view, the waste gas generation from some plant sections does not have to be zero. As long as the generated waste gas could be reused by the other sections to the maximum extent, the whole system emissions would be minimized. Therefore, to synchronize and balance operations of different sections of the entire chemical plant is the optimal way to reduce air emissions.

This paper will use two industrial emission examples to demonstrate the proposed idea. One is about the flare minimization for a chemical plant turnaround operations; the other is about air emission reductions for a petroleum refinery.

See more of this Session: Process and Product Development for Sustainability
See more of this Group/Topical: Process Development Division

Lowering Material and Energy Usage During Purging Ultra-High-Purity Gas Distribution Systems

Junpin Yao, Department of Chemical and Environmental Engineering, University of Arizona, Tucson, AZ

Hao Wang, Department of Chemical and Environmental Engineering, University of Arizona, Tucson, AZ

Farhang Shadman, Department of Chemical and Environmental Engineering, University of Arizona, Tucson, AZ

Carl Geisert, Intel Corporation, Chandler, AZ

The timely dry down of ultra-high-purity (UHP) gas distribution systems that are used to deliver process gases to point-of-use has historically been a challenge in semiconductor manufacturing. New technologies with shrinking geometries and with increased sensitivities to defects and trace contaminants, particularly moisture, make the problem more challenging with every new technology cycle. The cost and competition driven need to shorten new facility qualification schedules is in direct conflict with the basic physics and time dependency of moisture elimination from new piping and distribution system components due to the facts that moisture molecules dissociatively chemisorb on the surface of gas delivery pipes producing a tightly bound layer with a high activation energy of desorption. Upset events such as a contaminated delivery or catastrophic failure of a purifier can shut down a manufacturing line for weeks while the distribution system is decontaminated and dried down. Less ultra-pure gas and energy usage in the dry down or cleanup process directly satisfies the requirements of environmentally friendly semiconductor manufacturing processes.

These challenges underscore the need for thorough characterization of the physics and chemistry involved in system dry down, as well as understanding the most effective methods for accelerating the dry down in a high volume manufacturing environment. In order to optimize the purge process with minimum purge time and/or total costs that include purge gas and energy usage, a dynamic process model that couples gas phase transport and the surface interactions between moisture molecules and the components of gas distribution systems was developed. The process model is capable of characterizing the interaction mechanisms and can be used to predict the dry down behavior of large scale industrial gas distribution systems on which it is mostly unfeasible to run tests that may disturb the regular high volume manufacturing processes. The model was verified through the combination of theoretical analysis and experiments on small scale systems. The unknown parameters, such as rate constants of adsorption and desorption and surface sorption density, in the model were estimated by fitting model predictions to experimental data. The process model is scalable and applicable to various size systems once the fluid dynamics properties of the system are fully characterized.

With the assistance of the process model, it is easy to run a parametric study on the impacts of purge gas flow rate, purity, temperature, and system pressure on the dry down process of a large industrial scale system. Furthermore, during purging and cleaning processes at system start up, system recovery from disturbance or during the normal operation of gas distribution, the process model essentially can be applied as a distribution simulator that can be used to develop optimum purging recipes for minimum purge time and/or costs for an industrial scale gas distribution system.

The results of both experiments and model simulation indicated that the cleanup process is sensitive to the inherent properties of the gas delivery pipes, such as surface roughness, exposure concentration, and purge conditions. As compared to the conventional purge method with high purity and high flow rate of purge gas without heating the system and changing the system pressure, the results showed that different combinations of purge gas purity and flow rate at different purge stages can lead to the reduction of 37%

of purge cost and 49% of ultra-pure purge gas while slight increase of purge time. On a small gas distribution system, zone heating and pressure cyclic purge were experimentally proven to be two effective methods to accelerate the dry down process. For instance, an elctropolished stainless steel pipe with 65 inch in length and 1.5 inch in O.D. that was originally exposed to 200 ppb of moisture at room temperature, the experimental data showed that ultra pure purge gas usage can be dramatically decreased by 55% and 35% by zone heating and cyclic purge, respectively, as compared to conventional purge. Model simulation showed further save on purge gas usage can be achieved on large scale systems by these two methods. Zone heating is a very straight forward approach to drive moisture out from short and straight gas delivery pipes, while pressure cyclic purge is a more practical method for a complex configuration system. In a pressure cyclic purge process, the number of cycles, the time duration of each cycle, the ratio of high pressure to low pressure and the rate of pressure change are the key factors affecting the effects of cyclic purge on the cleanup process.

The model developed in this study can be applied as a generic dry down simulator of large scale UHP gas distribution systems to seek optimum purge conditions with the purpose of reducing purge gas usage and energy and/or cleanup time.

See more of this Session: Process and Product Development for Sustainability
See more of this Group/Topical: Process Development Division

REACTIVE DISTILLATION IN A MIXED ACID ESTERIFICATION PROCESS FOR BIOBASED CHEMICALS PRODUCTION

Alvaro Orjuela[1,2], Aspi kolah[1], Xi Hong[1], Carl T. Lira[1], and Dennis J. Miller[1]
(1) Michigan State University, East Lansing, MI, USA
(2) Universidad Nacional de Colombia, Bogotá, Colombia

Introduction

Bio-based succinic acid (SA) obtained by fermentation is considered a major building block for a future sustainable chemical industry. It can be used as feedstock and intermediate in production of solvents, surfactants, biodegradable polymers, specialty chemicals and commodities [1-7]. Commonly, SA produced by fermentation is obtained as sodium, calcium or ammonium salt in diluted aqueous solutions. These solutions might contain acetic acid (AcAc), and also small amounts of pyruvic, lactic, and/or formic acid [7], making difficult to isolate pure SA crystals.

Among the different alternatives to recover carboxylic acids from culture media, esterification has proven successful. In the early 1920's, it was demonstrated that acids from fermentation could be retrieved by esterification with different alcohols [8]. Recovery of AcAc [9-20], lactic [21 - 24], citric [25], maleic [26], and even SA [1, 7, 26, 27, 28, 29] from diluted aqueous solutions using this process has been reported.

Esters as well as SA can be used as raw materials for succinate derivatives [7]. Specifically, diethyl succinate (DES) produced by reaction with ethanol (EtOH) can be used directly as a green solvent, additive in fragrance formulation, chemical intermediate, additive for biodiesel, and, if produced economically, as a future route to PBS polymers. SA esterification to DES proceeds sequentially through a series – parallel reaction scheme as shown in Figure 1.

Considering that the major contaminant of bio-based succinic acid is AcAc, when EtOH is used in mixed acids esterification, ethyl acetate (EtAc) can be recovered as a valuable byproduct. In traditional separations, other carboxylic acids have to be removed and are lost as waste streams, thus affecting the efficiency of the process.

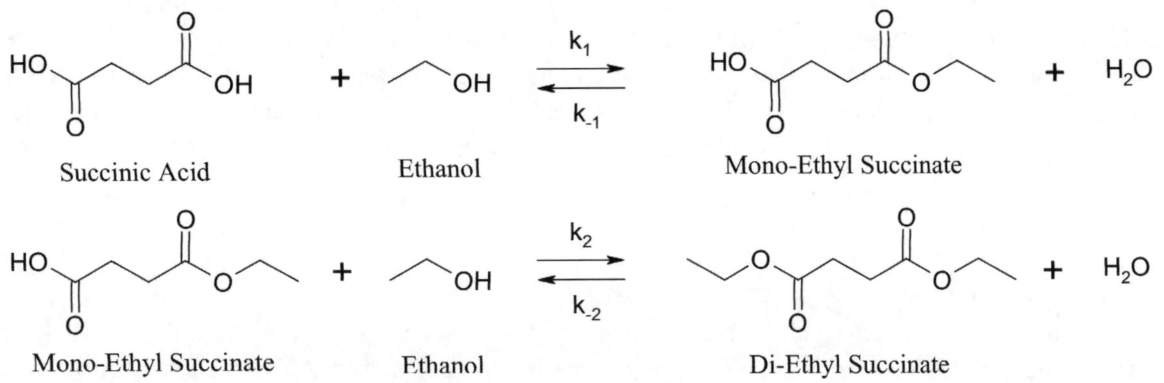

Figure 1. Esterification of succinic acid with ethanol

Because the extent of esterification is limited by chemical equilibrium, removal of products is required to achieve complete conversion. In this regard, reactive distillation (RD) provides a positive synergistic effect in reacting and separating the products simultaneously.

In many applications of RD, esters have either, the highest or the lowest volatility of the reacting mixture, allowing separation from water (H_2O). In the case of EtAc production, a hetero-azeotrope with water (H_2O) and EtOH is obtained, and further purification is required [30, 31]. Succinic acid esterification by RD does not fit in any of these categories, since DES has lower volatility than EtOH and H_2O, but higher than monoethyl succinate (MES) and SA which are essentially non-volatile. Therefore, obtaining pure DES is only possible if complete conversion is achieved in the reactive zone of the column.

There are few reports of RD applied to esterification of succinic species. Bart *et al* [32] studied methyl succinate production using succinic anhydride and methanol in a semi-batch reactive distillation cascade. Miller *et al* [27] illustrated the capability of a RD to produce DES. SA conversion of 45% was achieved in the column with a pre-reactor configuration. Sutton *et al* [26] described a RD process for maleic acid recovery by esterification with methanol, and they suggest that this technology could be used with similar success for succinic acid from fermentation broths. Bauduin *et al* [33] describe a novel process using a RD column to obtain dibutyl succinate from ammonium succinate and butanol. A dibutyl succinate yield of 96.5% is obtained, but the product is contaminated with nitrogen-containing byproducts.

In this work, a process to upgrade and separate mixed SA and AcAc by esterification with EtOH using a RD process is presented. A favorable RD configuration for high conversion of acids and high purity of DES has been established. Experimental results from a continuous pilot-plat RD column are presented and effects of important variables in RD are described.

Material and methods

Materials

Succinic acid (SA, 100 %, Sigma-Aldrich), ethanol (EtOH, 200 proof, Decon Labs), water (H_2O, HPLC grade, J.T. Baker), and acetic acid (AcAc, 99.9%, Aristar) were used in the present experiments without further purification. Purity was confirmed by gas chromatography. *n*-Butanol (99.9%, Mallinckrodt), and acetonitrile (HPLC grade, EMD) were used as internal standard and solvent during gas chromatographic analysis. Hydranal-coulomat E solution (Riedel-de Haën) was used in Karl-Fisher analysis for water concentration measurements.

Analysis

Samples were analyzed using both HPLC and GC. A Shimadzu® 2010 gas chromatograph, equipped with a Flame ionization detector and AOC-5000 autoinjector, was used. A 15 m Alltech® EC-WAX column (0.53 mm i.d., 1.20 μm film thickness) was used with the following temperature program: column initial temperature 313 K (1.37

min), ramp at 30 K/min to 353 K, and ramp at 40 K/min to 523 K (hold 4 min). The injection port was maintained at 553 K with a split ratio 5:1; detector temperature was 573 K. Hydrogen was used as carrier gas (50 cm/s), and volume injections of 1 μL were used.

A HPLC system with a Waters® 717 Autosampler, a Waters® 410 Refractive Index detector and a Perkin Elmer LC90 UV detector was used for SA and MES analysis. A 100 x 7.6 mm Bio-Rad® fast acid analysis column was used for the separation. The mobile phase consisted of a 5mM aqueous solution of H_2SO_4 at a flow rate of 1 ml/min. Since retention time of EtOH and MES was similar, only UV detection was used to evaluate MES.

Samples of known composition in the range of interest were used for calibration of GC and HPLC, and were injected in triplicate to obtain repeatability within 0.5% by mass. Karl-Fisher analysis (Aquacount coulometric titrator AQ-2100) was used to measure water content in reagents and samples.

Reactive distillation column description

Continuous RD experiments were carried out in a pilot-plant stainless steel column with an internal diameter of 5.1 cm and 488 cm total height. A sketch of the RD column is presented in Figure 2.

The column consists of six sections separated by liquid redistributors. The reactive zone (360 cm) consists of sections of structured packing Katapack-S® (Sulzer Chemtech Ltd) containing approximately 76 Kg/m^3 column volume of 0.3 mm diameter Amberlyst 70® cation exchange resin (Rohm and Haas Co.). The non-reactive enrichment (73.7 cm) and the stripping (41.7 cm) sections contain empty Katamax® structured packing (Koch-Glitsch Ltd.). The column and the reboiler are wrapped with electric heating tapes and glass wood bands to minimize heat loss. The tapes are individually controlled to minimize temperature differences with surroundings and thus minimize heat loss. Internal and external surface thermocouples are used to register temperature profiles in the column.

The column reboiler working volume is 1500 cm^3 and it is controlled with an overflow outlet. Heat is supplied with a 2.4 kW stainless steel coated electric resistance heater in contact with the liquid. The column is equipped with a total condenser with a chiller capable of reaching 253.15 K to recover diethyl ether (DEE) if produced. The vapor flow at the top is controlled with a solenoid valve maintaining constant pressure in the column. A constant liquid level reservoir is used to collect condensate, with an overflow outlet for distillate withdrawal. Reflux is also dispensed from this reservoir via a controlled flow pump; reflux flow rate is measured with a coriolis flowmeter.

Two pumps are used to feed solutions at controlled rates which are measured by weight change of feed tanks. Two top feed ports are located above the reactive zone. The upper one is used when the column operated without reflux. A third feed port is located below the reactive zone. Inlet streams were fed at the temperature of the feed stages by preheating them in traced and insulated pipelines. Ten temperature ports and four sampling ports are located along the column.

A 0.05 m^3 stirred glass reactor is used as pre-reactor, and is batch operated at total reflux for 24 h before feeding to the column. In the reactor, dry Amberlyst 15® cation exchange resin (Rohm and Haas Co.) is used as catalyst (1 % by total mass).

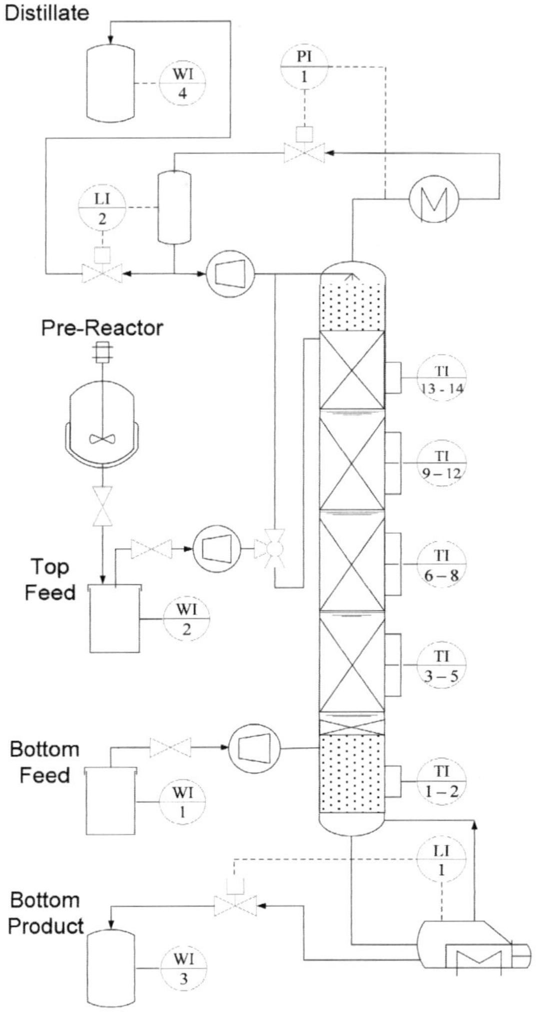

Figure 2. Scheme of pilot plant RD column with internal catalytic packing Katapack-S® (middle zone) and Katamax® structured packing (doted zone)

Operation of the column

Succinic acid is a crystalline solid at room temperature with limited solubility (~7 wt%) in EtOH and H_2O at 298 K. The succinic acid feed is introduced at the top of the RD column as a solution in EtOH at different concentrations or as pre-reacted mixture. An additional ethanol stream was fed at the bottom below the reactive zone.

In typical operation, the column is started by turning on the reboiler heater and the external heating tapes as the vapor reaches the different zones of the column. Total reflux is maintained until a stable temperature profile and pressure are reached. Then, feed pumps are started at the specified rates. Steady state is generally achieved after three reboiler volumes are collected from the bottom product stream. Inlet and outlet flows and temperature profiles are recorded during the run. The acid value (AV) of the bottom stream was measured every 30 min by titration with 0.1M solution of NaOH to evaluate the approach to steady state. Steady state samples were collected from distillate and bottom streams, and from sampling ports in the column.

Results and discussion

A list of experimental conditions evaluated is presented in Table 1. In all runs, a large molar excess of EtOH was used due to the low solubility of SA at room temperature (approx. 7 wt%). The feed rate and temperature of the acid stream was kept as constant as possible in all experiments to facilitate comparison. The effect of pressure, reflux ratio, ethanol flow rate (bottom feed), and acetic acid concentration were evaluated without pre-reaction. Runs 11 and 14 were performed using a pre-reactor - column configuration using SA/AcAc weight ratios of 3/1 and 10/1 in the acid feed. Higher concentrations of SA were allowed in those experiments because DES and MES are soluble in EtOH.

Table 1. Experimental conditions for RD

Run No.	Pre-reaction	Reboiler Power (kW)	P (psia)	Reflux ratio	Top Feed (Acid solution in EtOH)					Bottom feed (100% EtOH)		
					%w SA	%w AcAc	Flow (g/min)	Flow (mmol/min)	T (C)	Flow (g/min)	Flow (mmol/min)	T (C)
5B	No	0.72	45	0	6.67	-	30.233	629.56	90	-	-	-
5C	No	0.55	45	0	6.67	-	30.4	633.04	80	-	-	-
7A	No	0.744	20	0	6.67	-	29.764	619.79	86.7	-	-	-
8A	No	1.08	45	0	6.67	-	31.06	646.78	88	4.973	107.94	97.2
8B	No	0.98	45	0	6.67	-	31.22	650.02	88.2	10.33	224.28	88.5
9B	No	1.13	45	0.588	6.67	-	29.96	623.89	90	-	-	-
9C	No	1.16	45	1.880	6.67	-	30.75	640.38	90	-	-	-
11A	Yes†	1.152	45	0	25	-	30	551.90	79	-	-	-
11B	Yes†	1.536	45	0	25	-	31.65	582.25	82.5	16.6	360.32	79
12A	No	0.96	45	0	6.67	2.17	30.3	627.62	82	-	-	-
12C	No	1.392	45	0.363	6.67	2.17	30.3	627.62	80	11.2	243.43	81.4
14A	Yes†	1.632	45	0	25	2.5	31.23	570.65	90.5	16.55	359.26	88.6

† Acid concentration are reported before pre-reaction

An example of the outlet flow profiles obtained in the RD column during the approach to steady state is presented in Figure 3. The total time to reach steady state was 18 to 24 h depending on the bottom outlet flow. Scatter in flow data are due to pulsating behavior obtained by using overflow level control in the reboiler and the condensates reservoir. The summarized results of RD experiments are presented in Tables 2 and 3.

Table 2. Summary of results obtained in RD experiments

<table>
<tr><th rowspan="2">Run No.</th><th rowspan="2">Reboiler T (C)</th><th rowspan="2">Reboiler Power (kW)</th><th colspan="4">Conversion %</th><th colspan="2" rowspan="2">Selectivity to DES %</th><th colspan="2">Distillate</th><th colspan="2">Bottoms</th><th colspan="6">Recovery Percentage in outlet streams §</th></tr>
<tr><th colspan="2">SA</th><th colspan="2">AcAc</th><th>Flow g/min</th><th>Flow mmol/min</th><th>Flow g/min</th><th>Flow mmol/min</th><th>DES Bot.</th><th>MES Bot.</th><th>AcAc Top</th><th>EtAc Top</th><th>EtOH Top</th><th>H_2O Top</th></tr>
<tr><td>5B</td><td>120</td><td>0.72</td><td colspan="2">98.56</td><td colspan="2">-</td><td colspan="2">89.98</td><td>11.4</td><td>245.6</td><td>22.83</td><td>376.32</td><td>99.88</td><td>100</td><td>-</td><td>-</td><td>42.74</td><td>6.17</td></tr>
<tr><td>5C</td><td>120</td><td>0.55</td><td colspan="2">97.54</td><td colspan="2">-</td><td colspan="2">90.60</td><td>4.02</td><td>84.59</td><td>34.33</td><td>555.91</td><td>99.99</td><td>100</td><td>-</td><td>-</td><td>13.98</td><td>0.935</td></tr>
<tr><td>7A</td><td>105</td><td>0.744</td><td colspan="2">97.85</td><td colspan="2">-</td><td colspan="2">88.75</td><td>18.5</td><td>406.52</td><td>13.98</td><td>229.52</td><td>99.82</td><td>100</td><td>-</td><td>-</td><td>67.95</td><td>25.15</td></tr>
<tr><td>8A</td><td>120</td><td>1.08</td><td colspan="2">99.16</td><td colspan="2">-</td><td colspan="2">88.94</td><td>25.5</td><td>562.87</td><td>11.96</td><td>185.85</td><td>100</td><td>100</td><td>-</td><td>-</td><td>78.2</td><td>49.9</td></tr>
<tr><td>8B</td><td>120</td><td>0.98</td><td colspan="2">99.14</td><td colspan="2">-</td><td colspan="2">92.25</td><td>31.2</td><td>693.94</td><td>11.74</td><td>175.53</td><td>100</td><td>100</td><td>-</td><td>-</td><td>80.74</td><td>80.41</td></tr>
<tr><td>9B</td><td>120</td><td>1.13</td><td colspan="2">99.77</td><td colspan="2">-</td><td colspan="2">89.53</td><td>19.1</td><td>432.72</td><td>11.78</td><td>173.0</td><td>100</td><td>100</td><td>-</td><td>-</td><td>77.11</td><td>75.1</td></tr>
<tr><td>9C</td><td>121</td><td>1.16</td><td colspan="2">99.65</td><td colspan="2">-</td><td colspan="2">91.02</td><td>20.7</td><td>466.61</td><td>11.12</td><td>165.07</td><td>100</td><td>100</td><td>-</td><td>-</td><td>76.5</td><td>71.9</td></tr>
<tr><td>11A</td><td>135</td><td>1.152</td><td rowspan="2">92.19‡</td><td>66.89</td><td colspan="2">-</td><td rowspan="2">62.69‡</td><td>63.44</td><td>20</td><td>476.74</td><td>11.67</td><td>85.05</td><td>100</td><td>99.16</td><td>-</td><td>-</td><td>98.82</td><td>82.25</td></tr>
<tr><td>11B</td><td>178</td><td>1.536</td><td>86.66</td><td colspan="2">-</td><td>98.09</td><td>35.6</td><td>849.92</td><td>11.84</td><td>76.11</td><td>99.41</td><td>64.77</td><td>-</td><td>-</td><td>98.51</td><td>99.71</td></tr>
<tr><td>12A</td><td>121</td><td>0.96</td><td colspan="2">99.03</td><td colspan="2">95.0</td><td colspan="2">89.11</td><td>20.4</td><td>450.66</td><td>11.27</td><td>166.28</td><td>99.87</td><td>90.62</td><td>60.46</td><td>99.44</td><td>75.5</td><td>68.87</td></tr>
<tr><td>12C</td><td>121</td><td>1.392</td><td colspan="2">99.71</td><td colspan="2">93.33</td><td colspan="2">93.48</td><td>32.2</td><td>721.68</td><td>11.07</td><td>156.23</td><td>99.95</td><td>75.20</td><td>78.58</td><td>99.7</td><td>82.91</td><td>93.25</td></tr>
<tr><td>14A</td><td>231</td><td>1.632</td><td>95.34‡</td><td>90.02</td><td>82.1‡</td><td>47.77</td><td>60.88‡</td><td>98.46</td><td>71.55</td><td>862.98</td><td>11.5</td><td>83.17</td><td>100</td><td>92.45</td><td>83.17</td><td>99.99</td><td>99.24</td><td>99.73</td></tr>
</table>

‡ Data evaluated on pre-reactor product before processing in RD column.

§ Recovery percentage calculated with reference to overall outlet flow of each component.

Table 3. Summary of results obtained in RD experiments

Run No.	Distillate composition % w								Bottoms Composition % w							
	H_2O	DEE	EtAc	EtOH	AcAc	MES	DES	SA	H_2O	DEE	EtAc	EtOH	AcAc	MES	DES	SA
5B	0.41	3.05	-	96.00	-	0.00	0.03	0.06	3.81	0.02	-	79.5	-	1.42	15.18	0.12
5C	0.21	9.07	-	90.72	-	0.00	0.00	0.00	3.39	0.02	-	85	-	0.96	11	0.19
7A	0.86	0.29	-	98.74	-	0.00	0.02	0.09	3.99	0	-	72.86	-	2.21	20.72	0.22
8A	1.25	1.11	-	97.64	-	0.00	0.00	0.00	3.18	0.00	-	68.79	-	2.63	25.22	0.17
8B	1.79	1.05	-	97.17	-	0.00	0.00	0.00	1.36	0	-	72.64	-	1.7	24.11	0.18
9B	3.12	1.65	-	95.23	-	0.00	0.00	0.00	2.37	0	-	64.64	-	2.94	30	0.05
9C	2.87	2.17	-	94.97	-	0.00	0.00	0.00	2.55	0.01	-	66.4	-	2.37	28.6	0.08
11A	6.60	0.95	-	92.30	-	0.14	0.00	0.00	2.59	0	-	2	-	30.3	63.3	1.76
11B	7.06	1.58	-	90.99	-	0.18	0.18	0	0.06	0	-	4.161	-	1	94.07	0.7
12A	3.06	2.08	4.19	90.41	0.10	0.14	0.02	0.00	2.94	0.02	0.05	62.34	0.14	2.92	31.4	0.2
12C	3.34	1.31	2.64	92.46	0.11	0.14	0.00	0.00	0.82	0.03	0.03	65.06	0.10	1.43	32.47	0.06
14A	8.17	1.82	2.72	87.03	0.17	0.03	0.00	0.05	0.07	0.01	0	2.06	0.11	1.16	96.44	0.15

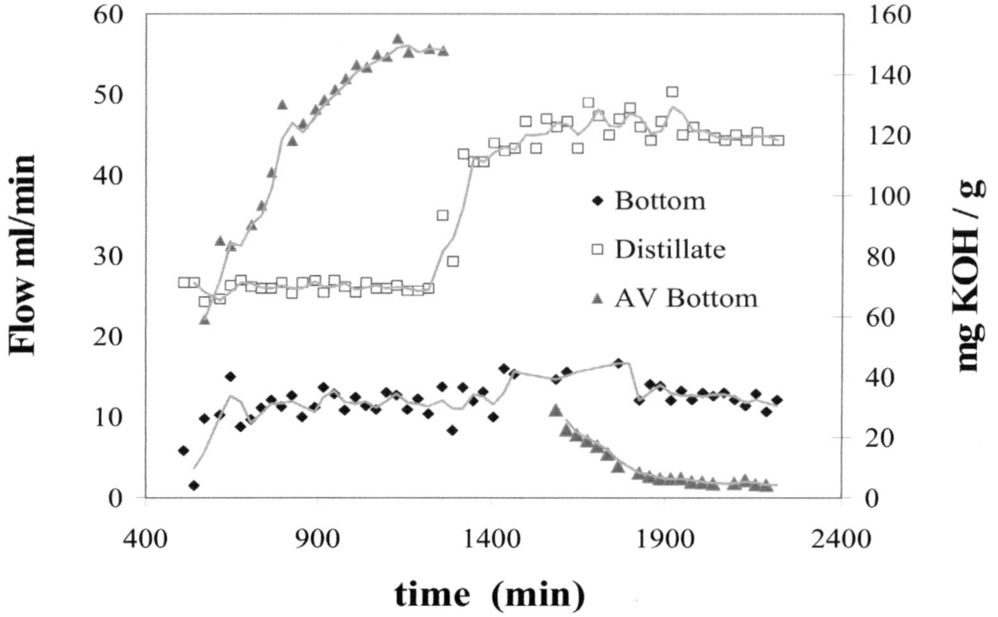

Figure 3. Dynamic behavior of outlet flows and bottoms acid value approaching steady state in Run 11 A (high AV) and 11B (low AV).

In general, high conversions of SA were achieved in all experiments. This was expected considering the large excess of EtOH fed to the RD column. The effect of boilup rate is observed by comparing Runs 5B and 5C. When the reboiler heating power was reduced, the EtOH vapor rate through the reactive zone was lower, thus reducing SA conversion. Hence, EtOH accumulates in the reboiler and, because of its long residence time at high temperature, it dehydrates to DEE (DEE is recovered by the top). At low boilup rate, H_2O also remains in lower sections of the column, limiting SA conversion.

Al lower pressure (run 7A) and the same boilup rate as Run 5B, temperatures decrease and there is a lower concentration of ethanol in the reboiler. As a result, DEE production is reduced. Despite the large excess of ethanol, conversion is slightly reduced because the rate of reaction is slower at low temperatures.

When pure EtOH is fed below the reactive zone (Runs 8A and 8B), conversion and selectivity to DES increase. Water is stripped out the bottoms by the increased flow of ethanol, leaving the column only in the distillate.

In Runs 9B and 9C, reflux was used instead of EtOH feed. In this case, the conversion remains high because of the large excess of EtOH in the reactive zone. The concentration of SA in the bottoms decreases, and concentration of both H_2O and DEE in distillate increase because of enhanced esterification and dehydration.

Conversion of SA achieved in the pre-reactor was 93-95%, and selectivity to DES was around 61%. AcAc conversion in the prereactor was 82%. Because of the higher concentrations of DES and MES from the prereaction, the bottoms temperature in Runs 11 and 14 bottoms was higher than in other experiments. When pure EtOH was not fed to the column, high concentrations of MES were observed in the bottoms. In contrast, DES was the major product in the bottoms when a pre-reacted mixture and pure EtOH were fed to the column.

When AcAc was fed to the column along with succinic acid, the EtAc produced was recovered almost completely in the distillate, while succinate species were recovered in bottoms. However, the high boilup rate in Runs 11, 12A and 14 and the absence of reflux led to some MES and DES exiting the column in the distillate stream, thus indicating the need to control boilup rate in optimized column operation. Also, in these mixed acid esterification experiments, some AcAc remained in the bottoms product. This indicates that high evaporation rate and high temperature in the lower part of the column help to concentrate succinic species but do not allow AcAc to reach the reactive zone to be totally converted.

These experiments show that separation of fermentation products as esters is possible by RD, but it is clear that optimum conditions to operate a mixed acids esterification were not achieved and further studies must be conducted in order to improve the technology.

Conclusions

RD experiments were conducted on the esterification of SA, and mixtures of SA and AcAc with EtOH. Steady state was obtained after 24 of continuous operation and was verified with AV measurements of the bottoms stream. In general, high conversion of SA was obtained and DES as the major product was recovered from the bottom of the column. Long residence times of EtOH and high temperatures are required for high conversion of SA and the highest concentrations of DES in bottom product, although these conditions also increase dehydration of EtOH to DEE. Complete conversion of AcAc was not obtained simultaneously with SA conversion, indicating that optimum conditions for both reactions must be further studied.

Acknowledgements

This work was supported by the Michigan Economic Development Corporation, 21st Century Jobs Fund Project #95096.

References

1. C. Delhomme, D. Weuster-Botza, F. E. Kühn. (2009). Succinic acid from renewable resources as a C_4 building-block chemical—a review of the catalytic possibilities in aqueous media. Green Chem. 11, 13–26

2. A. Cukalovic, C. V. Stevens. (2008). Feasibility of production methods for succinic acid derivatives: a marriage of renewable resources and chemical technology. Biofuels, Bioprod. Biorefin. 2, 505–529.

3. I. Bechthold, K. Bretz, S. Kabasci, R. Kopitzky, A. Springe. (2008). Succinic Acid: A New Platform Chemical for Biobased Polymers from Renewable Resources. Chem. Eng. Technol. 31 (5), 647–654.

4. B. Hermann, M. Patel. (2007). Today's and tomorrow's bio-based bulk chemicals from white biotechnology: A techno-economic analysis, Appl. Biochem. Biotechnol. 136, 361–388.

5. O. Wolf, M. Crank, M. Patel, F. Marscheider-Weidemann, J. Schleich, B. Hüsing, G. Angerer. (2005). Techno-economic feasibility of large-scale production of Bio-based polymers in Europe, European Science and Technology Observatory, EUR 22103 EN.

6. J. Zeikus, M. Jain, P. Elankovan. (1999). Biotechnology of succinic acid production and markets for derived industrial products, Appl. Microbiol Biotechnol. 51, 545–552.

7. S. Varadarajan, D. J. Miller. (1999). Catalytic Upgrading of Fermentation-Derived Organic Acids. Biotechnol. Prog. 15, 845–854.

8. A. A. Backhaus. (1921). Esters from fermented distillery waste. US Patent 1400852.

9. R. Neumann, Y. Sasoon. (1983). Recovery of dilute acetic acid by esterification in a packed chemorectification column. Ind. Eng. Chem. Process Des. Dev. 23, 4 654–659.

10. E. M. Jones. (1997). Process for the recovery of an organic acid from the manufacture of a cellulose ester. US Patent 5648529.

11. M. Scates. (1997). Recovery of acetic acid from dilute aqueous streams formed during a carbonylation process. US Patent 5599976.

12. Z. Xu, A. Afacan, K. Chuang. (1999). Removal of acetic acid from water by catalytic distillation. Part 1. Experimental studies, Can. J. Chem. Eng. 77, 676–681.

13. Z. Xu, A. Afacan, K. Chuang, (1999). Removal of acetic acid from water by catalytic distillation. Part 2. Modeling and simulation studies. Can. J. Chem. Eng. 77, 682–687.

14. B. Saha, S. P. Chopade, S. M. Mahajani. (2000). Recovery of dilute acetic acid through esterification in a reactive distillation column. Catal. Today. 60 (1-2), 147–157.

15. C. Bianchi. (2003). A new method to clean industrial water from acetic acid via esterification. Appl. Catal., B. 40, 93–99.

16. A. Singh, A. A. Tiwari, S. M. Mahajani, R. D. Gudi. (2006). Recovery of Acetic acid from Aqueous Solutions by Reactive Distillation. Ind. Eng. Chem. Res. 45 (6), 2017–2025.

17. W. Hung, I. Lai, Y.W. Chen, S. B. Hung, H. B. Huang, M. J. Lee, C. C. Yu. (2005). Process Chemistry and Design Alternatives for Converting Dilute Acetic Acid to Esters in Reactive Distillation. Ind. Eng. Chem. Res. 45, 1722–1733.

18. V. Ragaini, C. L. Bianci, C. Pirola and G. Carvoli (2006). Increasing the Value of Dilute Acetic Acid Streams through Esterification. Part I. Experimental Analysis of the Reaction Zone, Appl. Catal. B: Environ. 64, 66–71.

19. A. Singh, A. Tiwari, V. Bansal, R. D. Gudi, S. M. Mahajani. (2007). Recovery of Acetic Acid by Reactive Distillation: Parametric Study and Nonlinear Dynamic Effects. Ind. Eng. Chem. Res. 46 (26), 9196–9204.

20. A. Arpornwichanop, K. Koomsup, S. Assabumrungrat. (2008). Hybrid reactive distillation systems for n-butyl acetate production from dilute acetic acid. J. Ind. Eng. Chem. 14, 796–803.

21. R. Kumar, H. Nanavati, S. B. Noronha, S. M. Mahajani. (2006). A continuous process for the recovery of lactic acid by reactive distillation. J. Chem. Technol. Biotech. 81 (11), 1767–1777.

22. H. G. Joglekar, I. Rahman, S. Babu, B. D. Kulkarni, A. Joshi. (2006). Comparative assessment of downstream processing options for lactic acid. Separation and Purification Technology 52, 1–17.

23. X. Sun, Q. Wang, W. Zhao, H. Ma, K. Sakata. (2006). Extraction and purification of lactic acid from fermentation broth by esterification and hydrolysis method. Sep. Pur. Technol. 49, 43–48.

24. W. Zhao, X. Sun, Q. Wang, H. Ma, Y. Teng. (2009). Lactic acid recovery from fermentation broth of kitchen garbage by esterification and hydrolysis method. Biomass and Bioenergy. 33, 21–25.

25. A. Kolah, N. Asthana, D. Vu, C. T. Lira, D. J. Miller. (2008). Triethyl citrate synthesis by reactive distillation. Ind. Eng. Chem. Res. 47 (4), 1017–1025.

26. D. Sutton, G. Reed, G. Hiles. (2007). Process for the production of esters of mono-, di or polycarboxylic acids. US Patent 5599976.

27. D. Miller, N. Asthana, A. Kolah, D. Vu, C. Lira. (2006). Process for reactive esterification distillation. US Patent Application 20060252956.

28. V. Budarin, R. Luque, D. J. Macquarrie, J. H. Clark. (2007). Towards a Bio-Based Industry: Benign Catalytic Esterifications of Succinic Acid in the Presence of Water. Chem. Eur. J. 13, 6914–6919.

29. R. Luque, C. Lin, D. Chenyu, D. J. Macquarrie, A. Koutinas, R. Wang, C. Webb, J. H. Clark. (2009). Chemical transformations of succinic acid recovered from fermentation broths by a novel direct vacuum distillation-crystallization method. Green Chemistry. 11(2), 193–200.

30. M. Klöler, E.Y. Kenig, A. Górak, A. P. Markusse, G. Kwant, P. Moritz. (2004). Investigation of different column configurations for the ethyl acetate synthesis via reactive distillation. Chem. Eng. Process. 43, 791–801.

31. Y. T. Tang, H. Huang, I. Chein. (2003). Design of a Complete Ethyl Acetate Reactive Distillation System. J. Chem. Eng. Jpn. 36 (11), 1352–1363.

32. H. Bart, J. Reidetschlager. (1995). Distillation with chemical reaction and apparatus selection, Sep. Sci. Technol. 30 (1995) 1849-1865.

33. C. Bauduin, W. Fischer, R. Pinkos, and E. Scholten. (2009). Process for the synthesis of alkyl esters polycarboxylic acid and process their hydrogenation. US Patent Application 20090137825 A1.

Process Intensification of a Non-Phosgene Route for Isocyanate Manufacture

Nitin V. Tople, Chemical and Petroleum Engineering, University of Pittsburgh, Pittsburgh, PA

Eric J. Beckman, University of Pittsburgh, Pittsburgh, PA

Isocyanates are commonly used as monomers to make polymers, such as polyurethanes which find significant application in the manufacture of rigid and flexible foams. Polyurethanes are also used in the production of adhesives, elastomers, and coatings. Isocyanates are currently manufactured by the phosgenation route which involves reacting an amine with phosgene. This process requires handling of hazardous chemicals and is also environmentally malignant due to issues related to disposal of the hydrochloric acid, formed as a byproduct.

McGhee et al have previously worked on a route to manufacture isocyanates from primary amine, carbon dioxide, a nitrogenous base and an electrophilic dehydrating agent. However, due to the generation of huge amount of salt of the dehydrating agent and the base and due to the absence of an appropriate recycling strategy this route could not be commercialized.

We present a modification to this route by eliminating the use of the nitrogenous base and hence reducing the salt formed. A recycling strategy is proposed with the goal of making the process commercially viable. A primary amine in presence of low concentrations of carbon dioxide is converted to their corresponding carbamate salt. In presence of a strong dehydrating agent like trifluoroacetic anhydride, the carbamate is converted to isocyanate and salt of the primary amine and trifluoroacetic acid. In excess of carbon dioxide, the zwitterion intermediate of the primary amine is formed which on reaction with a dehydrating agent forms isocyanate without any salt as byproduct. The theoretical isocyanate to amine ratio is 1:2 with the carbamate route and is 1:1 with the zwitterion route. Solubility of CO_2 is high in polar solvents at very low temperatures and atmospheric pressure as well as at moderately high pressures and atmospheric temperatures. We have studied the effect of different solvents, dehydrating agents and varying temperature and pressures on the overall yield of isocyanates with the aim of maximizing the isocyanate content in the product mixture.

McGhee, W.D., Paster M.D., Riley D.P., Ruettimann K.W., Solodar J.A, Waldman T.E, U.S. Patent # 5,451,697, 1995

See more of this Session: Process and Product Development for Sustainability
See more of this Group/Topical: Process Development Division

GLOBAL OPTIMIZATION OF MASS AND PROPERTY INTEGRATION NETWORKS

José M. Ponce-Ortega, Facultad de Ingeniería Química, Universidad Michoacana de San Nicolás de Hidalgo, Morelia, Michoacán, México.
Fabricio Nápoles-Rivera, Instituto Tecnológico de Celaya, Celaya, México
Arturo Jiménez-Gutiérrez, Instituto Tecnológico de Celaya, Celaya, Guanajuato, México
Mahmoud El-Halwagi, Texas A&M University, College Station, TX, USA

Introduction

Mass integration strategies have been widely used to achieve simultaneously economic and environmental goals. However, because process and environmental constraints do not only depend on streams compositions but also on their properties, it is important to develop methodologies that consider this important issue. Shelley and El-Halwagi[1] proposed a component-less design method based on tracking functionalities and properties of complex mixtures, instead of tracking their individual constituents and showed the possibility to include this new concept into a mass integration framework that later El-Halwagi et al.[2] called property integration. This new approach allowed the characterization of streams in terms of their properties considering them as part of the optimization process.

Another limitation of most of the reported mass integration strategies is that they have considered process and environmental constraints separately, using as optimality criterion either the wastewater minimization or the fresh water consumption minimization. Ponce-Ortega et al.[3] found that for a direct recycle scheme the simultaneous consideration of process and environmental constraints reduces the total annual cost of the network compared to a sequential approach. Later, Ponce-Ortega et al.[4] proposed a simplified formulation for a recycle and reuse scheme in which mixing of streams is avoided, which eliminates non linear terms in the property balances. The main difference between these formulations is that the former considered a centralized treatment facility whereas the latter considered a distributed or in-plant treatment system. Galan and Grossmann[5] have also reported how the latter approach can reduce the flowrate to be treated in each unit, thus contributing to minimizing the treatment cost.

This work proposes a rigorous model for a mass and property integration network under a recycle and reuse scheme (see Figure 1) where an in-plant treatment system is included; hence, the final waste does not need further treatment to be disposed. Both environmental and process constraints are considered simultaneously based on properties such as toxicity, theoretical oxygen demand (ThOD), pH, density and viscosity. Flowrate constraints are also considered. The resulting model is a mixed integer non linear problem (MINLP) which is solved using the strategy proposed by Karuppiah and Grossmann[6] to yield near global optimal solutions.

Model Description and Formulation

The proposed model is based on the recycle and reuse configuration shown in Figure 1. Process sources can be segregated and sent to any of the treatment units. It is assumed that only one property can be treated in each unit. The source leaving a treatment unit can be segregated and sent either to any other unit to treat a different property, to a process unit or to the waste for its final disposal. Fresh sources can be used to complement process sources in order to satisfy the demands in process sinks.

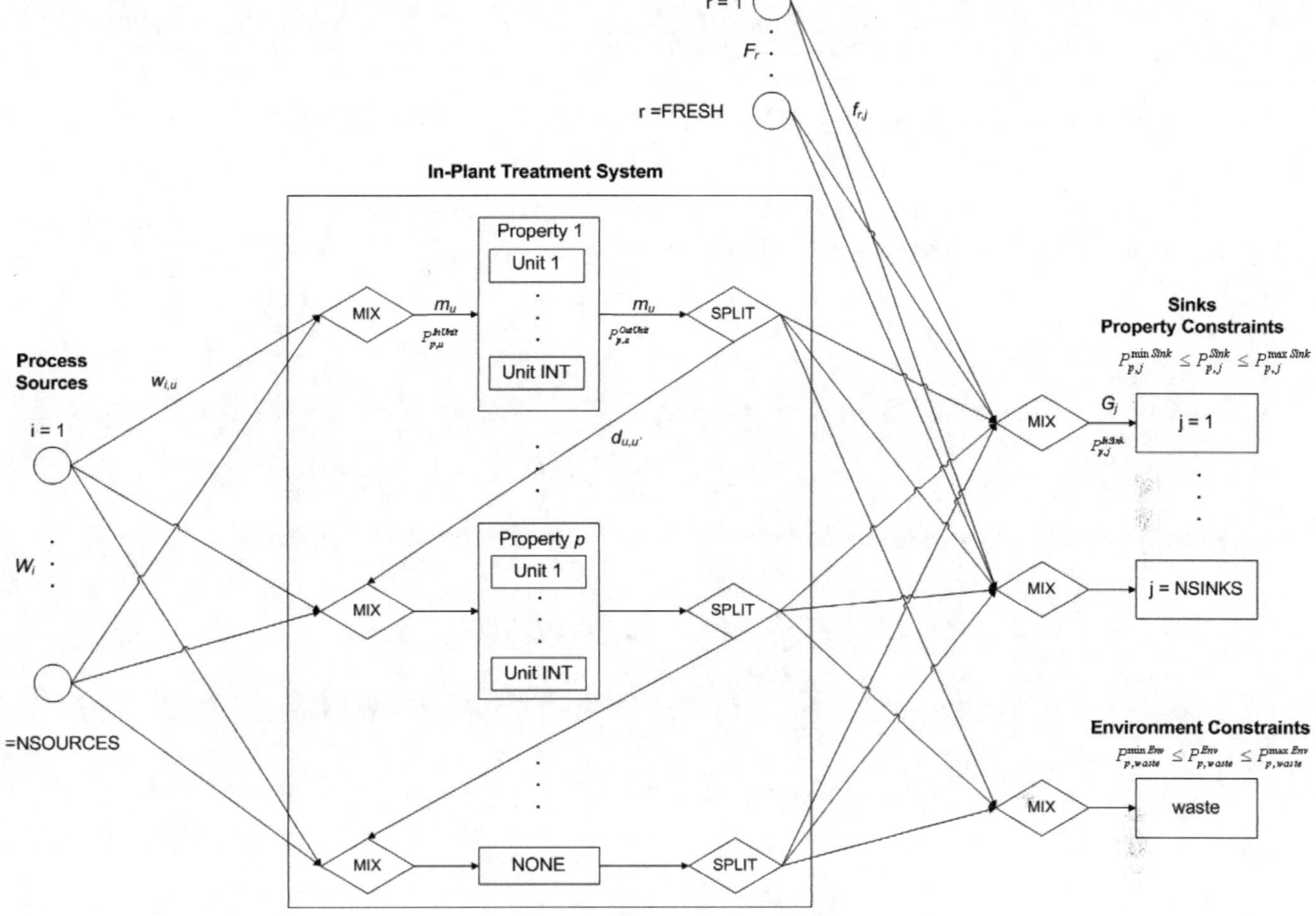

Figure 1. Recycle and Reuse Scheme

Model Formulation

Mass balances in the splitting and mixing points of the network:

$$F_r = \sum_{\substack{j \in NSINKS \\ j \neq waste}} f_{r,j}, \quad r \in FRESH$$

$$W_i = \sum_{u \in NINT} w_{i,u}, \quad i \in NSOURCES$$

$$\sum_{i \in NSOURCE} w_{i,u} + \sum_{\substack{u^1 \in NUNITS \\ u^1 \neq u}} d_{u^1,u}, = m_u \quad u \in NUNITS$$

$$m_u = \sum_{j \in SINKS} g_{u,j} + \sum_{\substack{u^1 \in NUNITS \\ u^1 \neq u}} d_{u,u^1}, \quad u \in NUNITS$$

$$G_j = \sum_{r \in FRESH} f_{r,j} + \sum_{u \in NUNITS} g_{u,j}, \quad j \in NSINKS$$

Property balances in the mixing points of the network:

$$\sum_{i \in NSOURCE} \left[w_{i,u} \psi_p \left(P_{p,i}^{InSource} \right) \right] + \sum_{\substack{u^1 \in NUNITS \\ u^1 \neq u}} \left[d_{u^1,u} \psi_p \left(P_{p,u^1}^{OutUnit} \right) \right] = m_u \psi_p \left(P_{p,u}^{InUnit} \right), \quad u \in NUNITS, p \in NPROP$$

$$G_j \psi_p \left(P_{p,j}^{InSink} \right) = \sum_{r \in NFRESH} \left[\psi_p \left(P_{p,r}^{InFresh} \right) f_{r,j} \right] + \sum_{u \in NUNITS} \left[g_{u,j} \psi_p \left(P_{p,u}^{OutUnit} \right) \right], j \in NSINKS, p \in NPROP$$

To treat any property, a set of units with known costs and correction factors are available. The following disjunction is used to choose which unit should be used for each property:

$$\underset{I(U) \in NUNITS}{\forall} \begin{bmatrix} Y_{I(U)}^{U(p)} \\ \psi_p \left(P_{p,U(p)}^{OutUnit} \right) = \psi_p \left(P_{p,U(p)}^{InUnit} \right) \left(1 - \alpha_{I(U)}^{U(p)} \right) \\ Cost_{U(p)}^{unit} = Cost_{I(U)}^{U(p)} m_{U(p)} \end{bmatrix}$$

A convex hull[7] reformulation is used to model the disjunction as follows:

$$\sum_{I(U)} y_{I(U)}^{U(p')} = 1, \quad U(p') \in NUNITS$$

$$\psi_{p'} \left(P_{p',U(p')}^{OutUnit} \right) = \sum_{I(U)} \psi_{p'} \left(p_{p',U(p')}^{OutUnit,I(U)} \right), \quad U(p') \in NUNITS,$$

$$\psi_{p'} \left(P_{p',U(p')}^{InUnit} \right) = \sum_{I(U)} \psi_{p'} \left(p_{p',U(p')}^{InUnit,I(U)} \right), \quad U(p') \in NUNITS$$

$$m_{U(p')} = \sum_{I(U)} m_{U(p')}^{I(U)}, \quad U(p') \in NUNITS$$

$$Cost_{U(p')}^{unit} = \sum_{I(U)} Cost_{U(p')}^{unit,I(U)}, \quad U(p') \in NUNITS$$

Notice that only one property can be treated in each unit and the other properties remain unchanged; therefore, the property balance for the properties not treated is:

$$\psi_{p'} \left(P_{p',U(p')}^{OutUnit} \right) = \psi_{p'} \left(P_{p',U(p')}^{InUnit} \right), \qquad U(p') \in NUNITS, \forall p \neq p'$$

Process and environmental constraints are considered simultaneously:

$$P_{p,j}^{\min Sink} \leq P_{p,j}^{inSink} \leq P_{p,j}^{\max Sink}, \quad j \in NSINK, j \neq waste, P \in NPROP$$

$$P_{p,j}^{\min Env} \leq P_{p,waste} \leq P_{p,j}^{\max Env}, \quad j = waste, P \in NPROP$$

The objective function consists in the minimization of the total annual cost for the network, which accounts for fresh sources, treatment units and piping costs as follows:

$$
\begin{aligned}
Min\,TAC = {} & H_Y \sum_{r\in FRESH} Cost_r^{Fresh} F_r \\
& + H_Y \sum_{u\in NUNITS} Cost_u^{Unit} \\
& + H_Y \sum_{u\in NUNITS} \sum_{i\in NSOURCES} \left(pip_{i,u}^{In} w_{i,u} \right) \\
& + H_Y \sum_{u\in NUNITS} \sum_{u^1\in NUNITS} \left(pip_{u,u^1}^{Int1} d_{u,u^1} \right) \\
& + H_Y \sum_{u\in NUNITS} \sum_{j\in NSINKS} \left(pip_{u,j}^{Exit} g_{u,j} \right) \\
& + H_Y \sum_{r\in NFRESH} \sum_{j\in NSINKS} \left(pip_{r,j}^{InFresh} f_{r,j} \right)
\end{aligned}
$$

Notice that the property balances contain several bilinear terms yielding a high non convex problem. The global optimization technique proposed by Karuppiah and Grossmann[6] is used to solve this problem.

Results and Discussions

Two cases of study with different dimensions were analyzed to show the applicability of the proposed model.

Example 1. The first example consists of two process sources, one fresh source and two process sinks. Table 1 shows the characteristics for the process and fresh sources.

Table 1. Process (W) and fresh (F) sources characteristics for Example 1

Stream	*Flowrate*	*Concentration component 1 (ppm)*	*Toxicity (%)*	*THOD (mg O2/l)*	*pH*	*Density*	*Viscosity*
W1	2,900	0.033	0.8	75	5.3	2	1.256
W2	2,450	0.022	0.5	88	5.1	2.208	1.22
F1	-	0	0	0	7	2.204	1.002

The global optimum for this problem is $205,908/year, and the optimal configuration is shown in Figure 2. It can be seen that every available unit has to be used to satisfy process and environmental constraints. The units used to treat the properties toxicity and pH are the ones with higher correction factors, whereas the rest of the units are the ones with lower correction factors and cost. Notice the existence of a fictitious unit called NONE; this unit is used exclusively for modeling purposes and its only operation is the mixing of streams. Both process and environmental constraints were satisfied, and it was found that the most stringent restrictions were satisfied near their extreme values. Table 2 shows a comparison between the simultaneous approach used here with respect to the sequential approach in which no environmental constraints where considered. It is worth noting that, in order to satisfy the environmental regulation, the waste obtained with the sequential approach needs an additional treatment system, yielding a total cost for the network of $271,806/year.

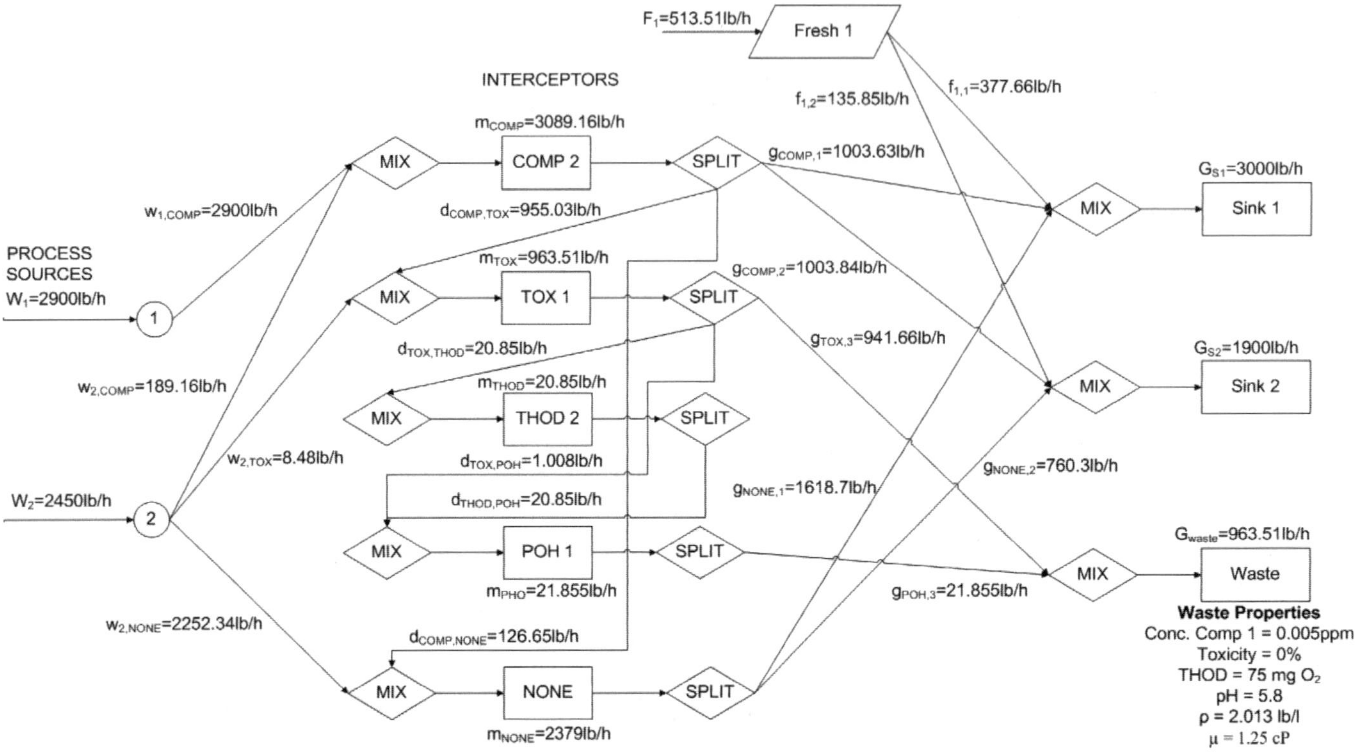

Figure 2. Optimal configuration for Example 1

Table 2. Comparison between simultaneous and sequential approach for Example 1

Concept	*Optimal Solution (Simultaneous Approach)*	*Sequential Approach*
F_1 (lb/hr)	513.51	515.39
Waste (lb/hr)	963.51	965.39
Fresh sources cost ($/year)	36,973	37,108
Treatment Units Cost ($/year)	158,764	72,611
Piping Cost ($/year)	10,170	9,939
Network Cost ($/year)	205,908	119,660
Waste Treatment System ($/year)	-	152,146
Total Annual Cost ($/year)	205,908	271,806

Example 2. This example contains six process sources, six process sinks and three fresh sources. Process and fresh sources data are given in Table 3.

The optimal solution has a total annual cost of $487,746/year. All the constraints are satisfied and as in the previous example, the properties were near their maximum or minimum values. The mass exchange network is shown in Figure 3.Table 4 gives some relevant results for the sequential and simultaneous approach. It can be seen that the simultaneous approach requires considerably less amount of fresh sources, and consequently generates a lower discharge to the environment. Additionally, the waste obtained in the sequential approach requires an additional treatment system to satisfy environmental constraints, and the total cost of the network including the aforementioned treatment system rises to $751,146/year.

Table 3. Process (W) and fresh (F) sources characteristics for Example 2

Stream	*Flowrate*	*Concentration component 1 (ppm)*	*Toxicity (%)*	*THOD (mg O2/l)*	*pH*	*Density*	*Viscosity*
W1	3100	0.16	0.4	95	4.4	2.1	1.236
W2	1800	0.1	0.7	85	3.9	2.208	1.220
W3	1750	0.11	1.3	90	4.7	2.305	1.241
W4	2000	0.12	0.8	100	4.7	2.105	1.256
W5	1300	0.09	0.4	100	3.8	2.305	1.260
W6	1400	0.2	1.5	100	5.7	2.102	1.259
F1	-	0	0	0	7	2.204	1.002
F2	-	0.01	0.1	0.01	6.8	2.209	0.992
F3	-	0.09	0.5	0.00	7.1	2.215	0.988

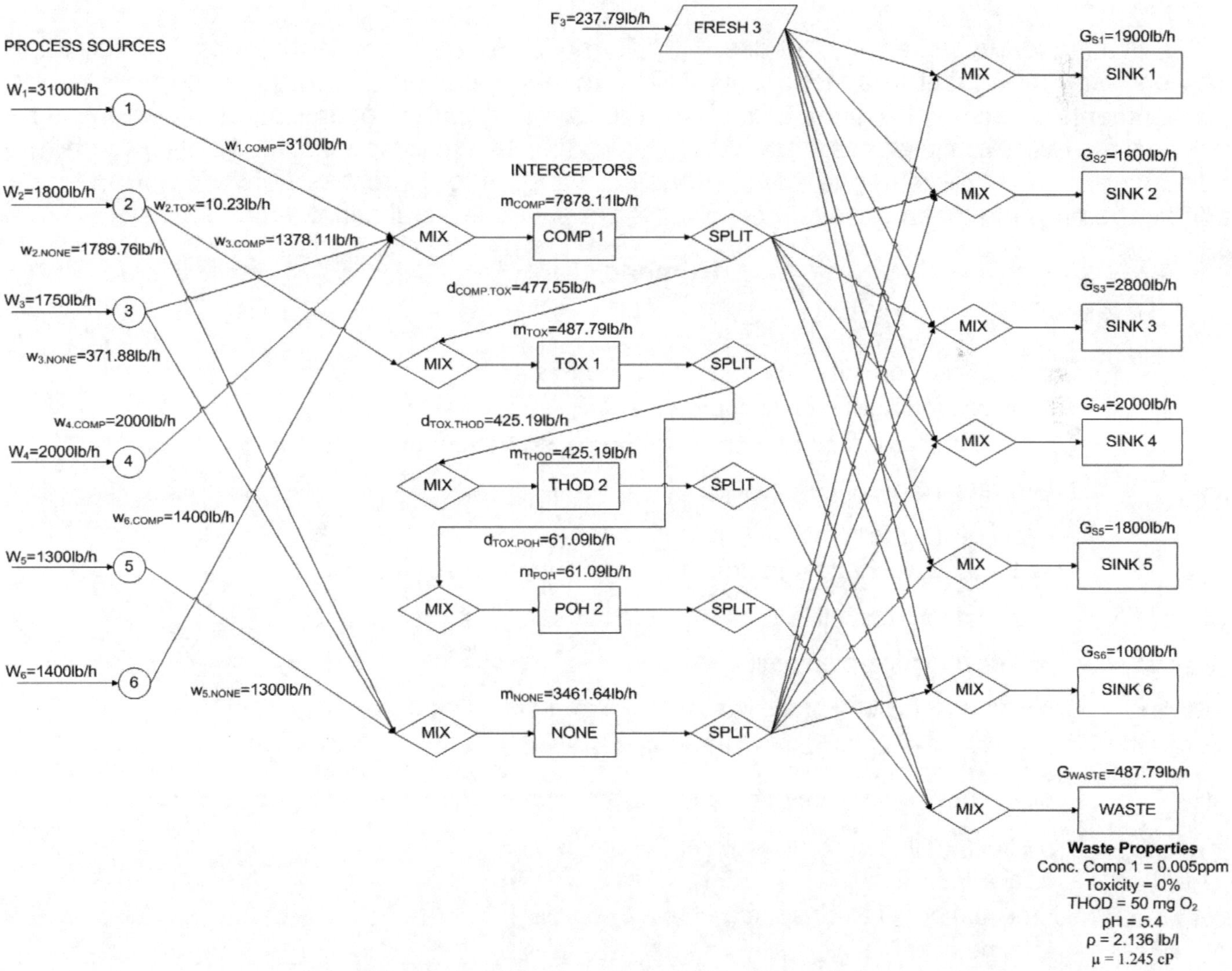

Figure 3. Optimal configuration for Example 2

Table 4. Comparison between simultaneous and sequential approach for Example 2

Concept	*Optimal Solution (Simultaneous Approach)*	*Sequential Approach*
F_1 (lb/hr)	-	1800
F_3 (lb/hr)	237.7	166
Waste (lb/hr)	487.7	2216
Fresh Sources Cost ($/year)	5,707	119,200
Treatment Units Cost ($/year)	460,451	205,123
Piping Cost ($/year)	21,587	20,792
Network Cost ($/year)	487,746	345,116
Waste Treatment System ($/year)	-	406,030
Total Annual Cost ($/year)	487,746	751,146

Conclusions

We have presented a rigorous MINLP model for a mass and property integration network with an in-plant treatment system. The model considers simultaneously process and environmental constraints, and it was solved using a global optimization algorithm. The solution for two examples has shown the application of the proposed methodology, and its advantage over a typical sequential approach of integrating a process network first and then adding the property treatment systems to meet the environmental constraints.

Nomenclature

F_r	total flow rate for fresh source *r*
$f_{r,j}$	segregated flow rate from fresh source *r* to sink *j*
W_i	total flow rate for process source *i*
$w_{i,u}$	segregated flow rate from process source *i* to treatment unit *u*
$d_{u^1,u}$	flow rate into treatment unit *u*
G_j	total flow rate for process sink *j*
$g_{u,j}$	segregated flow rate from treatment unit *u* to process sink *j*
$Cost_r^{Fresh}$	unit cost for fresh utility *r*
$Cost_u^{Unit}$	unit cost for treatment unit *u*
$pip_{i,u}$	piping cost to send process source *i* to treatment unit *u*
$pip_{u^1,u}$	piping cost to send segregated flow rate from treatment unit u^1 to treatment unit *u*
$pip_{u,j}$	piping cost to send segregated flow rate from treatment unit *u* to process sink *j*
H_Y	plant operating time, hr per year
TAC	total annual cost
N_{Fresh}	total number of fresh sources
N_{Sinks}	total number of sinks
$N_{Sources}$	total number of process sources

FRESH	set for fresh sources, $\{r \mid r=1,\ldots,N_{Fresh}\}$
NSINKS	set for sinks, $\{j \mid j=1,\ldots,N_{Sinks}\}$
NSOURCES	set for process sources, $\{i \mid i=1,\ldots,N_{Sources}\}$
z	composition
Tox	toxicity
THOD	Theoretical oxygen demand
pH	potential of hydrogen
waste	total flow rate for the waste stream discharged to the environment

Greek symbols

$\alpha_{I(u)}^{u(p)}$	efficiency of property interceptor for property *p*
ψ_p	property operator for the mixing rule for property *p*
ρ	density
μ	viscosity

Indices

i	process source
In	inlet
j	sink
max	maximum
min	minimum
Out	out
p	property
r	fresh source
Sink	sink
Source	source
u	*unit*
waste	waste discharged to the environment

References

1. Shelley, M. D.; El-Halwagi, M. M., Component-Less Design of Recovery and Allocation Systems: A Functionality-Based Clustering Approach. *Computers & Chemical Engineering.* **2000,** 24, 2081-2091.
2. El-Halwagi, M. M.; Glaswog, I. M.; Qin, X.; Eden, M. R., Property Integration: Componentless Design Techniques and Visualization Tools. *AIChE Journal.* **2004,** 50 (8), 1854-1869.
3. Ponce-Ortega, J. M.; Hortua, A. C.; El-Halwagi, M. M.; Jiménez Gutiérrez, A., A Property-Based Optimization of Direct Recycle Networks and Wastewater Treatment Processes. *AIChE Journal.* **2009,** 55 (9), 2329-2344.
4. Ponce-Ortega, J. M.; Jiménez Gutiérrez, A.; El-Halwagi, M. M., Simplified Formulation for the Mass and Property Interception Networks Considering Environmental Constraints. *Submited to Computers and chemical Engineering.* **2009**.
5. Galan, B.; Grossmann, I., Optimal Design of Distributed Wastewater Treatment Networks. *Industrial and Engineering Chemistry Research.* **1998**, (37), 4036-4048.

6. Karuppiah, R.; Grossmann, I., Global Optimization for the Synthesis of Integrated Water Systems in Chemical Processes. *Computers & Chemical Engineering.* **2006,** 20 (4), 650-673.
7. Raman, R.; Grossmann, I., Modeling and Computational Techniques for Logic Based Integer Programming. *Computers & Chemical Engineering.* **1994,** 18 (7), 563-578.

Solar-Thermal Water Splitting Process Development and Economics to Produce Renewable H2

Melinda M. Channel, Chemical & Biological Engineering, University of Colorado, Boulder, CO

Jonathan Scheffe, Chemical & Biological Engineering, University of Colorado, Boulder, CO

Alan W. Weimer, Chemical & Biological Engineering, University of Colorado, Boulder, CO

Using concentrated solar energy from a heliostat field, metal ferrite materials (MFe_2O_4) can be reduced according to the reaction $MFe_2O_4 = 1.2\ FeO + 0.4Fe_2O_3 + MO + 0.3O_2$ at 1,400°C. The resulting solid solution is then used to split water at 1,000°C via an exothermic oxidation reaction $1.2FeO + 0.4Fe_2O_3 + MO + 0.6H_2O = MFe_2O_4 + 0.6H_2$, producing high purity renewable hydrogen. Various metals are under investigation including M = Co, Ni, and Zn. This process has been designed and modeled using AspenPlus™ to produced 100,000 kg H2/day with reduction conversions of 35%, 70%, and 100%. Once the base case AspenPlus™ models were developed, the process was optimized to produce electricity by generating steam in the heat removal steps of the cycle. The models were then used to size and cost all capital equipment and the solar heliostat field to determine the total capital investment (TCI) of the process. The TCI and fixed/variable operating costs were then entered into the Department of Energy's (DOE) H2A economic analysis program to establish the allowable purchase price of metal ferrite for a given hydrogen selling price to meet the DOE cost targets through 2017. The results of this analysis are very promising with solar to receiver annual average efficiencies of 45% and thermal cycle efficiencies as high as 50%. The targeted hydrogen selling price of $4/kg was met with reasonable allowable ferrite capital costs. The economic analysis indicates that the development of solar-thermal water splitting processes using ferrite materials in a redox cycle is a potentially economical pathway to renewable hydrogen.

See more of this Session: Process and Product Development for Sustainability II
See more of this Group/Topical: Process Development Division

Mechanisms of Aqueous Extraction of Soybean Oil

K. A. Campbell, C. E. Glatz
Department of Chemical and Biological Engineering
Iowa State University, Ames, IA 50011

A paper presented at the AIChE Annual Meeting, Nashville, TN, November 2009

Introduction

Aqueous extraction processing (AEP) of soy is a promising green alternative to hexane extraction processing. While recent advances in AEP techniques have increased the recovery of free oil to 85% [1], AEP yields are still less than typical yields from industrial hexane extraction processes. AEP uses water as an extraction medium, dissolving soluble cellular materials and allowing the release of oil into the bulk liquid phase, from which the oil can be recovered by centrifugation resulting in a cream emulsion which can be broken to recover free oil [1-3]. Approximately 10-15% of the oil released from the solid fraction also remains in the aqueous fraction as an emulsion stable toward creaming [1]. Because of the immiscible nature of the oil/water system, the poorly understood mechanisms of oil release are intrinsically different than those from hexane extraction processes. In order to increase yields of AEP, a thorough understanding of the extraction mechanisms is needed.

Materials and Methods

Soybeans were prepared at the Iowa State University Center for Crops Utilization Research using local common variety, 2005 and 2007 harvest. Four modes of comminution were investigated: flour, flake, flour from flake, and extrudate. Flour was prepared by passing cracked dehulled soybeans twice through a pin-mill. Soy flake was prepared by passing cracked dehulled soybeans through a smooth surface roller to a thickness of approximately 0.25 mm. Flour from flake was produced by passing flakes once through a pin-mill. Extrudate was produced by extruding soy flake as described previously [1]. All flours were stored at -20 °C until use. Protex 7L (P7L), a neutral metalo-endoprotease from *Bacillus amyloliquefaciens,* with optimum temperature and pH ranges of 40-60 °C and 6-8, respectively, and Protex 6L (P6L), an alkaline serine-endoprotease from *Bacillus lichenformis*, optimum temperature and pH ranges 30-70 °C, and 7-10, respectively, were kindly provided by Genencor International (Rochester, NY) as a water-propylene glycol solution containing <10% protein (w/w) as received from the manufacturer.

Flour from flake particle size distribution

Flour from flake particle size distribution was measured by laser light scattering (Mastersizer 2000 S, Malvern Instruments, Ltd., Chicago, IL). A flour suspension was prepared by adding 1 g flour to 175 mL de-ionized water initially at pH 2.7 for a final pH of 4.5, then agitating 3 hr to break up clumps. Aliquots were introduced to the instrument by transfer pipette.

Microscopy

For transmission electron microscopy (TEM) and light microscopy (LM), samples were fixed and embedded following Bair and Snyder [4] with minor modifications at the Microscopy and NanoImaging Facility at Iowa State University. Whole seed was imbibed overnight in deionized water; comminuted samples were placed directly into fixative solution.

Extraction

Extractions were performed in a 2 L jacketed glass reaction vessel (Model 4742, Chemglass Inc., Vineland, NJ) held at 50 °C by a circulating water bath, and agitated by a stirrer (Model BDC 3030, Caframo, Ltd., Wiarton, Ontario) with a 2-inch diameter, 3-bladed screw impeller. Constant pH of 8.0 (flour from flake) and 9.0 (extrudate) was maintained using an autotitrator (Model 718 Stat Titrino, Metrohm, Ltd., Herisau, Switzerland). For response surface extractions flour from flake was extracted in de-ionized water without enzyme with an agitation rate of 500 rpm for 120 min to extract as much oil as possible without enzyme. After 120 min, enzyme was added and agitation rate was changed to the test condition for an additional 120 min. All other extractions were carried out for four hours with any additions made at the start of the extraction. Extractions for flour from flake and extrudate utilized P7L and P6L, respectively. At the end of the extraction, two 35-ml samples were withdrawn by siphon from the center of the reactor. Samples were centrifuged at 3,000 x g for 15 min. Sample fractions (solid and liquid) were massed and solid fractions were retained for analysis. Solid fraction moisture content was determined by drying in an oven at 130 °C for 12-15 hours. Solid fraction oil content was determined on freeze-dried samples using a Goldfisch extraction apparatus (AOCS Official Method Bc 3-49). Oil and protein yields were determined based on the difference between content of the starting material and the content of the solid fraction. Protein dissolution was calculated as including the protein in the liquid fraction plus that solubilzed protein entrained in the solid fraction, which was estimated by multiplying the liquid fraction protein concentration times the mass of water in the solid fraction.

Interfacial Tension

Interfacial tension of a soy oil/skim interface was measured by the Wilhelmy Plate method using a FACE Automatic Surface Tensiometer (Kyowa Interface Science Co. Ltd., Niiza-City, Japan). Skim was carefully poured over oil phase (Hy-Vee brand 100% soy oil, Hy-Vee, Inc., West Des Moines, Iowa) in a glass dish and allowed to stand one hour to achieve interfacial equilibrium before measurement.

Degree of Hydrolysis

Degree of hydrolysis (*DH*) was estimated using the pH stat technique [5] by measuring the volume of base added to maintain constant pH according to

$$DH = \frac{VN_b}{\alpha Ph_t} \tag{1}$$

where V and N_b are the volume and normality of base added, respectively, α is the fraction of deprotonated terminal protein residues (0.88 for soy protein at pH 8 and 50 °C [5]), P is the mass of protein being hydrolyzed, and h_t is the total number of peptide bonds per mass of protein (7.8 meqv/g soy protein [5]).

Results

Microscopic analysis of effects of comminution and extraction

Typical soy cotyledon cells are about 30-50 μm in diameter and 70-80 μm long [6-7]. Figure 1 shows a TEM of a native soybean cotyledon radial cross section. Most evident in the cross-section are the protein bodies, where about 80% of the soy protein is stored. The oil is stored in oil bodies, protein-phospholipid delimited lipid storage organelles [8] which fill most of the cytoplasmic network.

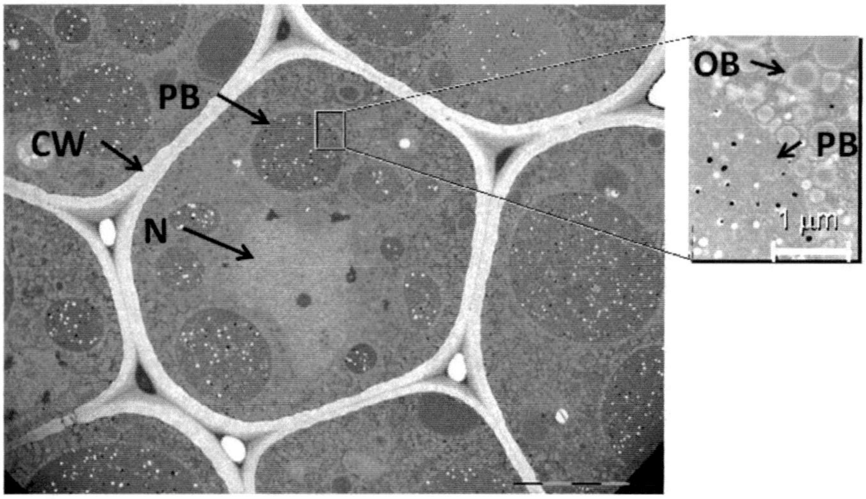

Figure 1. TEM of soybean cotyledon cell cross-section. PB, protein body; CW, cell wall; N, cell nucleus; OB, oil body.

Microstructure of soybeans comminuted by four alternative methods - milling, flaking, flaking followed by milling (flour from flake), and extrusion - were studied by microscopy. Images (Figure 2) were selected to show the range of cellular disruption observed. Regions of intact cells were found in all samples but the extruded material, which provided practically complete cellular disruption (Figure 2D). Ultrastructure of disrupted cells, shown by TEM (Figure 3), showed a range of oil body alterations. In intact cells, oil bodies were mostly unaltered from their pre-treatment appearance. In disrupted cells, some coalescence of oil bodies was observed, but many oil bodies remained intact even in completely disrupted cells.

Images of material after two hours of AEP both with and without protease for flour from flake and extruded material are shown in Figure 4. In flour from flake without protease (Figure 4A), very little of the residual material is extracellular. Protein bodies of disrupted cells have been dissolved, while large droplets of coalesced oil are present in some cells. Structural features of cells near the center of the flour from flake particles are unaffected by the extraction. Images of extruded material (Figure 4B) show oil droplets within a solid matrix both before and after extraction, although the amount of oil is visibly less after extraction. The oil not extracted after two hours of AEP of extruded soy is contained within the matrix material.

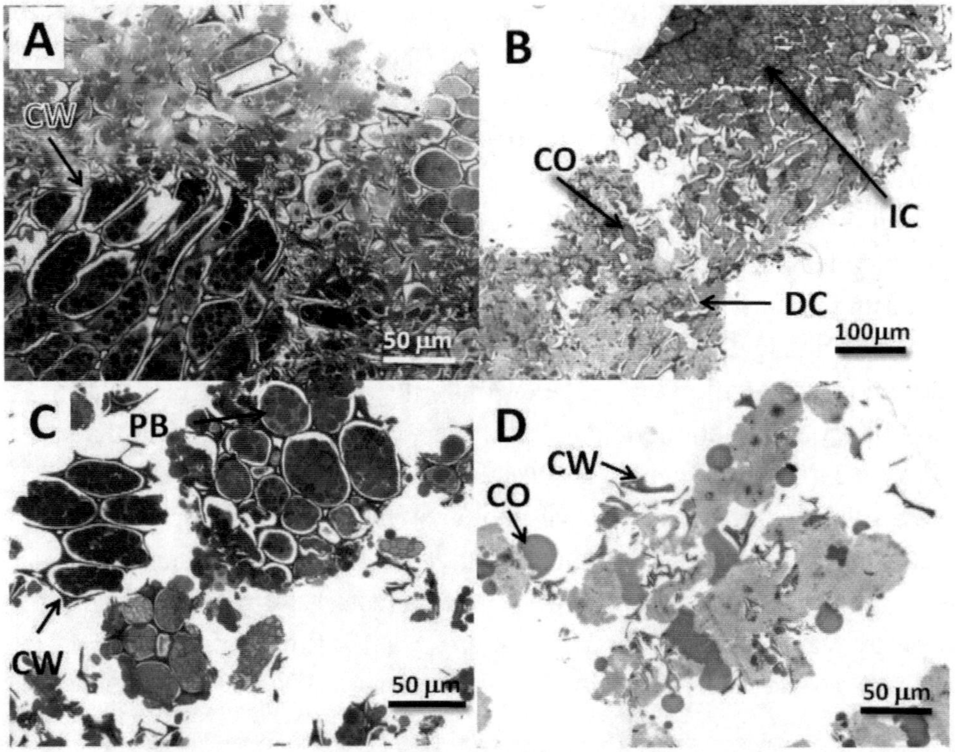

Figure 2. Light microscopy images of soy after various modes of comminution: (A), milling (flour); (B), flaking; (C), flaking followed by milling (flour from flake); (D), extrusion. Letters denote: PB, protein body; CO, coalesced oil; CW, cell wall fragment; IC, region of intact cells; DC, region of disrupted cells.

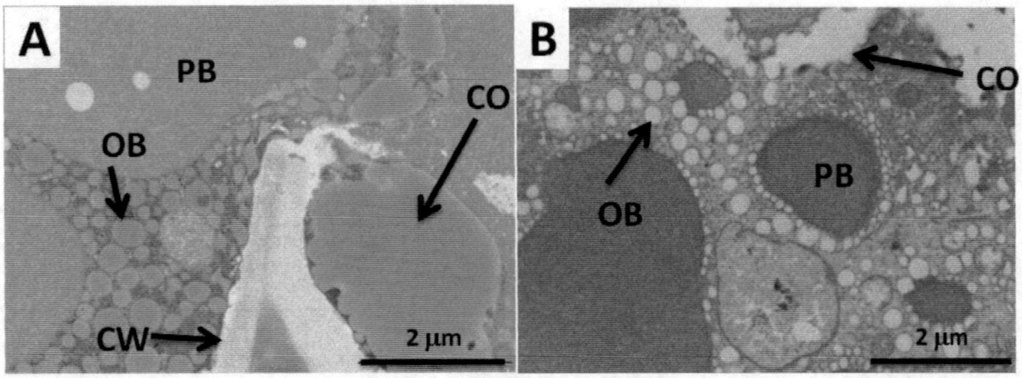

Figure 3. TEM of ultrastructure of (A) flour and (B) flour from flake before extraction. Letters denote: PB, protein body; OB, oil body; CO, coalesced oil; CW, cell wall. Osmium tetroxide stain likely did not penetrate the sample of image B, leaving the lipids with a transparent appearance.

For flour from flake, after two additional hours of extraction with 0.5% (w/w solid) P7L, oil extraction yield increased from 75% (+/- 1%) to 79% (+/- 1%). There was no noticeable change in the appearance of the residual flour from flake from that of the protease-free extraction (Figure 4C), possibly because only a small change in yield was achieved. It is not obvious based on these images why the addition of protease results in the increased release of oil. In the extruded material, however, the effect of protease is more pronounced, increasing yield from 68% to between 88% and 96% [1, 9]. Likewise, the extracted residual is drastically altered by the addition of protease (Figure 4D). Rather than a matrix enclosing oil droplets, as seen in Figure 4B, the protease-treated residual is loose and amorphous with little entrained oil, bearing little resemblance to the starting material. Because the increase in oil extraction is accompanied by an increase in dissolved protein [1], the images suggest that the proteolysis dissolves the matrix of insoluble denatured proteins, allowing the release of entrained oil. The fact that cellulase enzymes had no effect on extraction yields of extruded soy [9] confirms complete cellular disruption and a release mechanism aided by dissolution of the denatured protein. The residual should contain cell wall remnants, but these did not show up in the fraction of material sampled. There was a coarse material observed on the bottom of the centrifuge tube, so it is likely that after being freed from the matrix, the cell walls settled fastest during centrifugation. Samples taken for microscopy were taken mostly from the upper half of the residual bed in the centrifuge tubes.

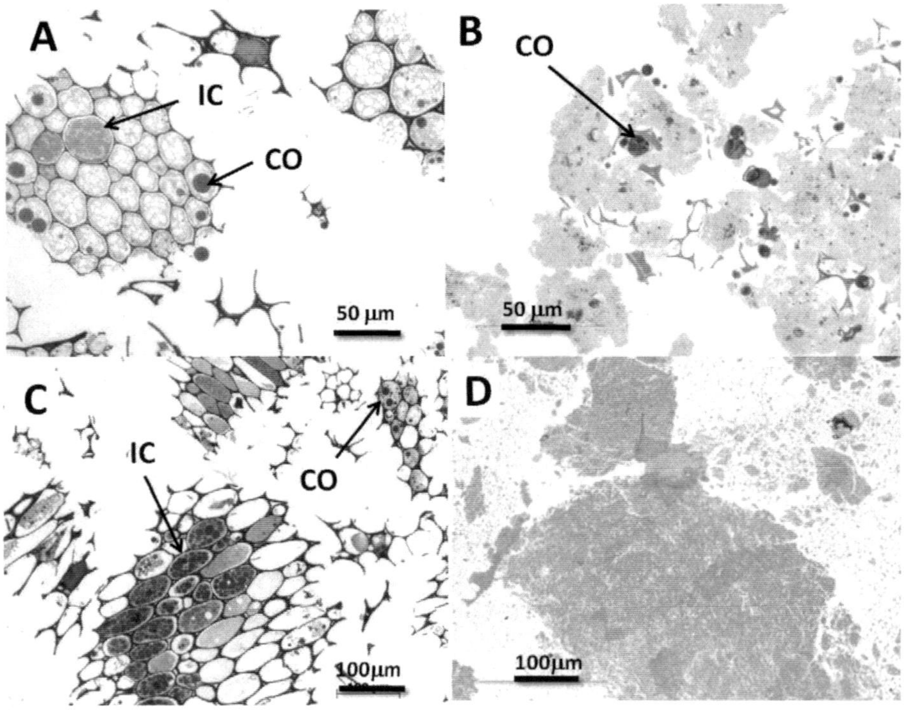

Figure 4. LM images of residual material after 2 hours of aqueous extraction. (A) Flour from flake without protease ; (B) extruded flake without protease; (C) flour from flake after two

hours with 0.5 % (w/w solid) Protex 7L; (D) extruded flakes after two hours with 0.5 % (w/w solid) Protex 6L . Letters denote: CO, coalesced oil; IC, region of intact cells.

Alteration of interfacial properties for extraction from flour from flake

Agitation is known to affect AEP extraction yield [10] which indicates that emulsification may be an important extraction mechanism. In protein-stabilized emulsions, droplet size is influenced by the protein-protein interactions resulting in the formation of viscoelastic interfacial films [11]. Treatments intended to vary the interfacial composition affect extraction yield as shown in Table 1. Adding 3% (w/w solid) SDS increased yields by 13% compared to extraction with no additions for flour from flake. This yield increase coincided with a modest reduction in interfacial surface tension from 5.3 mN/m (+/-0.2) without SDS to 2.1 mN/m (+/- 0.1) with SDS. If the formation of a viscoelastic interfacial film by disulfide bridging or other protein-protein interactions impede oil release from flour from flake, the addition of 3% (w/w solid) SDS appears to be successful in disrupting such a film.

Table1. Oil extraction yield, skim-oil interfacial tensions, and droplet size estimate for 4 hr extraction, 500 rpm with SDS, enzyme, and no additions using flour from flake.

Condition	Oil Extraction Yield (%)	Interfacial Tension of Skim-oil Interface (mN/m)
Oil-water	NA	12.3 A
No enzyme, S/L = 0.10	71.7 (+/-1)*	5.3 B,C
No enzyme, S/L = 0.15	71**	4.2 B
No enzyme, S/L = 0.05	83.0**	5.3 B,C
E/S = 0.5%, S/L = 0.10	81.0**	5.3 B,C
E/S = 1.9%, S/L = 0.10	81.7**	5.8 C
3% (w/w) SDS, S/L = 0.10	84.8 (+/- 0.2)***	2.1 D

*95% confidence interval for multiple experimental trials. **Final value of kinetic data. ***Mean +/- range for two replicate trials. Letters denote significant differences (0.05 level). NA- not applicable

While it is possible that proteolysis may increase yields by creating small polypeptides that are better emulsifiers than native proteins, this is not supported by interfacial surface tension data. The interfacial tension of skims extracted with or without protease (ca. 5 mN/m) were not significantly different (Table 1), suggesting a more complicated relationship between surface tension and extraction. If proteins adsorbed to the oil-water interface inhibit droplet break-up by forming a viscoelastic film, then a probable mechanism through which protease affects yield is by disrupting or preventing the formation of such a protein film.

Data in Table 1 show that the solid-liquid ratio has no significant effect on interfacial tension, although this parameter has a strong influence on yield. If the extraction yield (and therefore droplet mobility) is an indicator of intracellular oil droplet size, then the static interfacial tension does not give an accurate indication of droplet size in the absence of protease or SDS, as would be expected if viscoelastic effects were inhibiting droplet breakup. The fact that solid-liquid ratio has such a strong effect on extraction yield may support the hypothesis of protein adsorption hindering droplet breakup, thereby reducing extraction yield, as increasing bulk protein concentration would also increase adsorbed protein concentration.

Mechanistic model for AEP of flour from flake

Based on the microscopy of flour from flake and these emulsification hypotheses, several non-sequential events can be deduced to occur during aqueous extraction of flour: 1) instantaneous release of oil already outside of the cellular matrix from completely disrupted cells; 2) dissolution of protein bodies; 3) coalescence of oil bodies; 4) protein adsorption to the oil-water interface; 5) formation of a viscoelastic protein film at the oil-water interface; 6) break-up of coalesced droplets by inertial turbulent forces; 7) movement of droplets from the cellular matrix into the bulk fluid. Figure 5 shows a model constructed around these steps, assuming that steps (1) and (2) occur quickly. In the model, $P_{e,i}$ is oil from cells of a high degree of disruption, which is already outside of the cellular matrix and is readily removed. P_1 is small oil droplets (or intact oil bodies) from disrupted cells of the cellular matrix that can pass into the bulk medium after the protein bodies have dissolved. While this is occurring, oil in P_1 coalesces into P_2, a pool of larger droplets that have a lower mobility and are more difficult to extract. At longer times, a protein film forms around droplets in P_2, such that the turbulent forces are no longer capable of breaking them into droplets of an extractable size and this oil joins an unextractable pool, P_u. Oil in undisrupted cells, $P_{u,i}$, remains unextractable throughout the extraction. All kinetic processes are assumed to be first order. The total extracted oil, $P_{e,t}$ is the sum of the contributions from $P_{e,i}$ and P_e. The governing rate expressions are

$$\frac{dP_e}{dt} = k_1P_1 \quad k_3P_2 \tag{2}$$

$$-\frac{dP_1}{dt} = (k_1 + k_2)P_1 \tag{3}$$

$$-\frac{dP_2}{dt} = -k_2P_1 + (k_3 + k_4)P_2 \tag{4}$$

with initial conditions of $P_e = 0$, $P_1 = P_{1,o}$, $P_2 = 0$, and $P_u = 0$ and the overall mass balance

$$1 = P_{e,i} + P_e + P_1 + P_2 + P_u + P_{u,i} \tag{5}$$

Solving these gives the expression for total oil extraction

$$P_{e,t} = P_{e,i} + P_{1,o}\left[\left(-\frac{k_1}{K_1} - \frac{k_2k_3}{K_1(K_2-K_1)}\right)\left(e^{-K_1t} - 1\right) + \frac{k_2k_3}{K_2(K_2-K_1)}\left(e^{-K_2t} - 1\right)\right] \tag{6}$$

where $K_1 = k_1 + k_2$, and $K_2 = k_3 + k_4$.

The sizes of pools $P_{e,i}$, $P_{1,o}$, and $P_{u,i}$ can be estimated using particle size distribution data. $P_{u,i}$ represents completely intact cells, which, according to microscopy of flour from flake, must reside at the center of flour particles large enough to contain intact cells. The diameter of the intact core of a flour particle will be some average length smaller than the particle itself. The fraction of total volume occupied by intact cells will therefore be

$$I = \sum_{i=1}^{n} F_i \left(\frac{V_{I,i}}{V_i}\right) \tag{7}$$

for all $V_{I,i}$ greater than the average cell volume, where I is the volume fraction occupied by intact cells, F_i is the volume fraction of particles of size i as determined from light scattering analysis, $V_{I,i}$ is the volume of the intact core of particles of size i, and V_i is the volume of particles of size i. The average dimensional length of a soy cell is approximately 55 μm [4]. Assuming spherical geometry for flour particles, the intact cellular core of a flour particle must have a radius about 55 μm smaller than the particle radius, and therefore

$$V_{I,i} = \frac{4}{3}\pi(r_i - 55\ \mu\text{m})^3 \tag{8}$$

where r_i is the radius of particle size i. Using the flour from flake particle size distribution with Equations 7 and 8 gives a total intact cellular fractional volume of about 4.3% of total flour volume. This agrees very well with protein extraction data with this material. Average protein extraction yields ranged from 93% (+/- 0.5%) without protease to 95% (+/-0.7%) with protease, indicating that no more than 5% of the cells remained intact. Likewise, the fraction of oil in cells of a high degree of disruption, i.e. oil in particles less than the average cellular dimension, can be found with the cumulative size distribution function, shown in Figure 6. About 57% of total flour volume is made up of particles smaller than 55 μm, suggesting that the size of pool P_o should be near 0.57. By mass balance, the fraction of total oil in cells of a partial disruption, i.e. $P_{1,o}$, must be near 0.38. Because the pools $P_{e,i}$ and $P_{1,o}$ are independent of any extraction condition (N.B. the pools would depend on the type of comminution), there are only four variable fitting parameters (k_1, k_2, k_3, and k_4) to use to fit kinetic data for all extraction conditions.

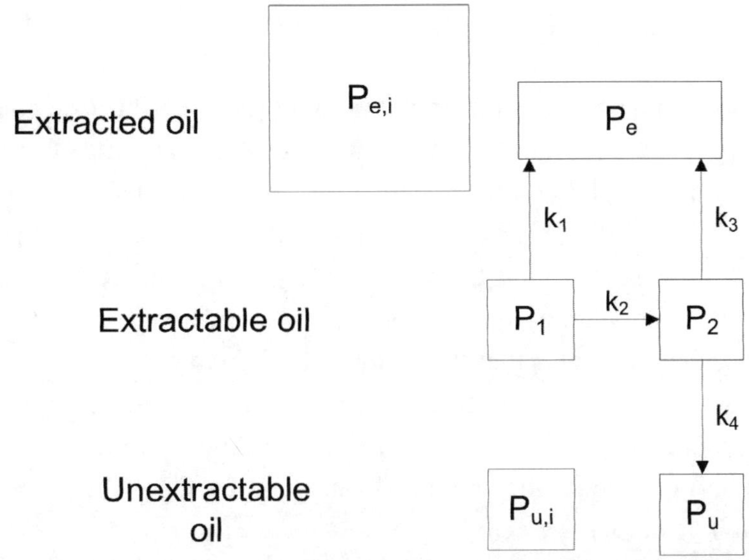

Figure 5. Oil extraction compartmental kinetic model for flour from flake. $P_{e,i}$ is oil from completely disrupted cells; P_1 is a pool of small oil droplets extracted quickly, P_2 is coalesced oil that is extracted slowly; P_u is unextractable oil; $P_{u,i}$ is unextractable oil contained within intact

cells. All of the oil is initially contained within compartments $P_{e,i}$, P_1, and $P_{u,i}$. As extraction progresses, oil in P_1 is either extracted or coalesces into large oil droplets, represented by P_2. Larger oil droplets may be released by emulsification, however the formation of a viscoelastic protein film around oil droplets prevents this and renders the oil droplets unextractable, represented by compartment P_u.

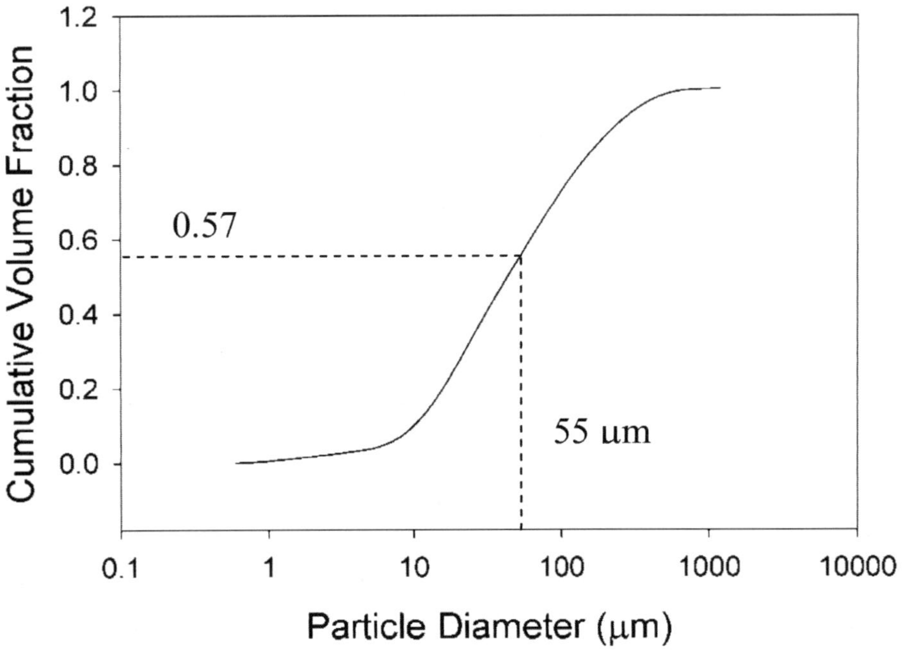

Figure 6. Cumulative particle size distribution (by volume) of flour from flake particles. Value shown is the volume fraction smaller than the average cellular size ($P_{e,i}$) of 55 μm.

Equation 6 and initial pool sizes from the particle size distribution gave good fits to kinetic extraction data, shown in Figures 7 and 8, with resulting parameter values shown in Table 2. In all cases, the value of k_1, the kinetic parameter for small droplet extraction, was at least an order of magnitude greater than k_3, the parameter for coalesced oil extraction, consistent with the model concepts where coalesced oil has a lower mobility within the cellular matrix than small droplets. For extraction without enzyme, values for k_2, k_3, and k_4 were roughly invariant with extraction conditions. This indicates that the shapes of the extraction curves are approximately the same after the first five minutes of extraction, regardless of solid-liquid ratio, and that the final yield is determined mostly by the rate of oil release in the first minutes of extraction, represented by the k_1 term. Indeed, the value of k_1 is enough greater for the S/L = 0.05 condition that the conversion to P_2 is less significant.

Table 2. Parameter estimates of non-linear regression fits of Equation 6 to oil extraction kinetic data for various solid to liquid ratios (S/L) enzyme to solid ratios (E/S) for extraction from flour from flake. Values for $P_{e,i}$ and $P_{1,o}$ were 0.57 and 0.38, respectively.

Parameter	S/L = 0.05 (no enzyme)	S/L = 0.10 (no enzyme)	S/L = 0.10 Rep (no enzyme)	S/L = 0.15 (no enzyme)	S/L = 0.15 Rep (no enzyme)	S/L = 0.19 (no enzyme)	S/L = 0.10 (E/S = 0.5%)	S/L = 0.10 (E/S = 1.9%)
k_1	0.331	0.059	0.101	0.018	0.022	0.064	0.170	0.500
k_2	0.290	0.307	0.361	0.042	0.103	0.233	0.258	0.611
k_3	0.003	0.006	0.005	0.001	0.004	0.006	0.029	0.046
k_4	0.006	0.019	0.014	0.006	0.014	0.025	0.044	0.092

With one exception, the addition of enzyme increased the rates of all the processes in this model. These effects can be explained in terms of the proteolytic action. Hydrolysis may increase the rate of release of small oil droplets (k_1) by either decreasing liquid viscosity or by increasing the rate of dissolution of protein bodies, as discussed above. However, Figure 8 shows that the protease has the largest effect on extraction at short extraction times (less than 30 min) while the degree of hydrolysis is still very low (less than 2%). This may indicate that the effect of hydrolysis is to create small polypeptide fragments that behave like LMW surfactants, such as SDS, and assist extraction at short extraction times. Large polypeptides may have a stronger influence on droplet interfacial behavior at longer extraction times after droplets have undergone multiple cycles of break-up and coalescence, leading to similar interfacial tension measurements seen in Table 1. Figure 8 also illustrates an advantage of EAEP of soy flours over soy extrudates in that increases in yields can be achieved with a minimization of protein alteration, leaving open the possibility of creating soy protein isolates with similar functional properties as conventional isolates.

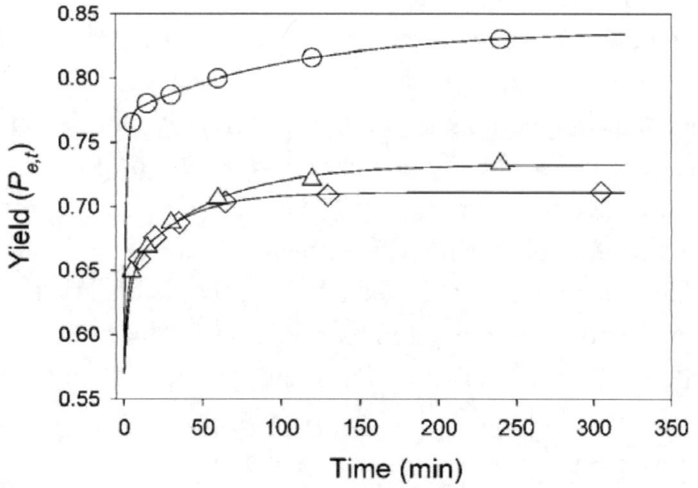

Figure 7. Kinetics of oil extraction for flour from flake with model fit for three different solid-liquid ratios. All extractions were without enzyme, with an agitation rate of 500 rpm. Curves represent model (Equation 6) using parameters in Table 2. Circles, S/L = 0.05; triangles, S/L = 0.10; diamonds, S/L = 0.19.

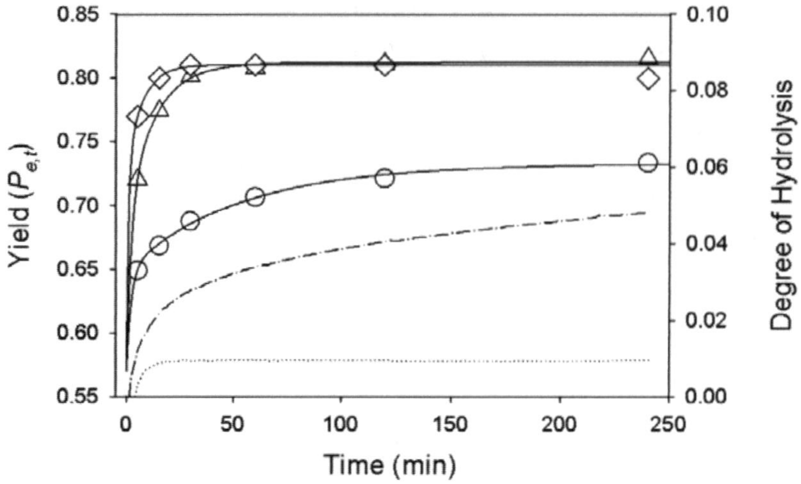

Figure 8. Kinetics of oil extraction yield and degree of hydrolysis for flour from flake with and without protease at a solid-liquid ratio of 0.10, and an agitation rate of 500 rpm. Solid curves represent model (Equation 6) using parameters in Table 2. Circles, no enzyme; triangles, E/S = 0.5%; diamonds, E/S = 1.9%; dotted curve, degree of hydrolysis for E/S = 0.5%, dotted-dashed curve, degree of hydrolysis for E/S = 1.9%

In terms of the other mechanisms discussed, hydrolysis may increase the rate of droplet coalescence (k_2) by disrupting the oleosin membrane of the oil bodies. The release of larger coalesced oil droplets (k_3) may also be increased if those droplets are more easily disrupted again into smaller droplets (i.e. they have a smaller maximum stable droplet diameter), which may occur if hydrolyzed proteins form weaker interfacial films. Finally, the formation of these films (k_4) may also occur more quickly with protein hydrolysates because smaller polypeptides would be able to diffuse more quickly to the interface than native proteins, and, because of the denaturation that occurs as a result of hydrolysis, no additional conformational changes would be necessary for intermolecular interactions to occur between polypeptides. It was also hypothesized above that the effect of enzyme may be to disrupt or prevent the formation of interfacial protein films, a state represented by P_u. However, forcing a very small value of k_4 in the model resulted in noticeably poorer fits (data not shown).

In summary, the extraction model based on microscopic observation, inferences of the interfacial properties, and pool sizes predicted by particle size distribution analysis fit the experimental data well. Observed initial extraction rates are consistent with diffusion-limited extraction rates. Hydrolysis data indicate that the role of proteolysis in flour from flake extraction may be to create small polypeptides that behave similar to a LMW surfactant.

Acknowledgements

This work was sponsored by the USDA CREES Grant #2005-34432-1406 and 2006-34432-17128, with additional funding from the Iowa State University Plant Science Institute. We would like to thank Tracey Pepper and Randall DenAdel for the microscopy, Prof. Tong Wang for use of her surface tensiometer, and Peter Bierschbach of Genencor, Intl, Julia Nobrega de

Moura, and the Iowa State University Center for Crops Utilization Research for providing materials used for this research.

References

1. Moura, J., et al., *Enzyme-assisted aqueous extraction of oil and protein from soybeans and cream de-emulsification.* Journal of the American Oil Chemists Society, 2008. **85**(10): p. 985-995.
2. Chabrand, R.M., et al., *Destabilization of the emulsion formed during aqueous extraction of soybean oil.* Journal of the American Oil Chemists Society, 2008. **85**(4): p. 383-390.
3. Chabrand, R.M. and C.E. Glatz, *Destabilization of the emulsion formed during the enzyme-assisted aqueous extraction of oil from soybean flour.* Enzyme and Microbial Technology, 2009. **45**(1): p. 28-35.
4. Bair, C.W. and H.E. Snyder, *Electron-Microscopy of Soybean Lipid Bodies.* Journal of the American Oil Chemists Society, 1980. **57**(9): p. 279-282.
5. Adler-Nissen, J., *Enzymatic hydrolysis of food proteins.* 1985, New York, NY: Elsevier Applied Science Publishers.
6. Wolf, W.J. and F.L. Baker, *Scanning Electron-Microscopy of Soybeans.* Cereal Science Today, 1972. **17**(5): p. 124-7.
7. Bair, C.W., *Microscopy of soybean seeds: Cellular and subcellular structure during germination, development, and processing with emphasis on lipid bodies.* 1979, Iowa State University: Ames, IA.
8. Tzen, J.T.C., et al., *Lipids, proteins, and structure of seed oil bodies from diverse species.* Plant phsysiology, 1993. **101**: p. 267-276.
9. Lamsal, B.P., P.A. Murphy, and L.A. Johnson, *Flaking and extrusion as mechanical treatments for enzyme-assisted aqueous extraction of oil from soybeans.* Journal of the American Oil Chemists Society, 2006. **83**(11): p. 973-979.
10. Rosenthal, A., D.L. Pyle, and K. Niranjan, *Simultaneous aqueous extraction of oil and protein from soybean: mechanisms for process and design.* Trans. Food and Bioproducts Processing, 1998. **76**: p. 224-230.
11. Mackie, A. and P. Wilde, *The role of interactions in defining the structure of mixed protein-surfactant interfaces.* Advances in Colloid and Interface Science, 2005. **117**(1-3): p. 3-13.

Thermodynamics of Bacteria to Determine Maximum Bioethanol Production

Craig A. Griffiths, School of Chemical and Metallurgical Engineering, Centre of Material and Process Synthesis (COMPS), University of the Witwatersrand, Johannesburg, South Africa

Diane Hildebrandt, School of Chemical and Metallurgical Engineering, Centre of Material and Process Synthesis (COMPS), University of the Witwatersrand, Johannesburg, South Africa

David Glasser, School of Chemical and Metallurgical Engineering, Centre of Material and Process Synthesis (COMPS), Johannesburg, South Africa

In a recent paper we have shown how one can calculate the Enthalpy and Gibbs Free Energy of formation of living cells. [Griffiths et al]. Once we have these values we are in principle in a position to treat biological processes in a similar way to normal chemical processes. [Patel et al, 2007]

As more emphasis is placed on sustainable process biological processes will become more important. Understanding how these processes work and what can be achieved from them will become valuable and necessary in plant design. These biological processes are inherently complicated, however they are not excluded from the laws of thermodynamics. We can treat the biological process like a typical chemical process. Using the thermodynamic properties of bacteria the heat (Enthalpy) and work (Gibbs Free Energy) flows of the process can be analysed.

By treating the system as a black box it can be simplified and the main equations can be set up. From the equations a feasible product spectrum can be calculated for the desired products. This data can be used to determine the maximum selectivity of the biological process and the by products that will be produced with it. The maximum selectivities can be used to evaluate the biological process's efficiency. The byproducts can help determine what impact the bioreactor would have on a large scale plant. In this paper we will look at an example of bioethanol production.

As the oil price gradually increases there is a greater push towards renewable fuels such as bioethanol. Bacteria used to produce bioethanol typically follow the Embden-Meyerhof pathway. The other products from this pathway are lactate and acetate. An important aspect of this process is how much ethanol we can produce relative to the lactate and acetate. This is the target of the research. The use of this target will show what side products must be made and their limits.

Further considerations are how this target depends on bacteria species, reactor design and process design. Using an overall systems approach, a thermodynamic analysis can be performed on the bacterial process. This analysis incorporates mass flow, heat flow (Enthalpy) and work flow (Gibbs Free Energy). This simple yet powerful investigation allows us to calculate the thermodynamic limit of ethanol production. The theoretical predictions are compared with data published in literature. Although this approach is illustrated on a bacterial process it can in principle be applied to any microorganisms such as yeast or fungus. This approach provides a powerful process synthesis tool that can be used to evaluate microbial process efficiency and process economics using minimum data.

Griffiths, C., Darko, G., Hildebrandt, D. & Glasser, D.., Thermodynamic Analysis of a Microbial Cell: A Process Synthesis Approach. Paper in preparation

Patel, B., Hildebrandt, D., Glasser, D. & Hausberger, B. Synthesis and Integration of Chemical Processes from a Mass, Energy, and Entropy Perspective. Ind. Eng. Chem. Res. **46**, 8756-8766 (2007).

See more of this Session: Process and Product Development for Sustainability II
See more of this Group/Topical: Process Development Division

MERCURY REMOVAL FROM COAL BY LEACHING WITH SULFURDIOXIDE GAS

AIChE 2009 Annual Meeting

Poornima Chateker, Department of Chemical Engineering, West Virginia University, Morgantown WV

Abstract

The objective of this research is to develop a standard method to reduce the mercury emissions from coal-fired power plants. This research focuses on pretreatment of coal to remove mercury prior to coal combustion. The purpose is to determine whether this pretreatment can replace the adsorption process of mercury vapor from the flue gas stream using activated carbon. The minute amount of mercury in Illinois No. 6 coal (0.216 ppm) was removed by flowing a gas stream containing 10% oxygen and 1000 ppm SO_2 into a coal-water slurry at 30 ml/s. Fifty grams of 35 – 65 mesh coal was leached each time in 500 ml solution. The temperature was varied from 50 to 80°C and the initial solution pH varied from 5.7 to 1.08. It was found that the mercury removal increased from 55.5% at 50°C to 99.2% at 75°C at the natural pH (5.7) and after 2 hours of reaction time. The pyrite removal percentages were much lower than those of mercury. Mercury removal was high enough to consider the application of this technology to a commercial process.

Introduction

Mercury from coal-fired utilities has been identified as one of the most hazardous air pollutants and the great potential public health concern. Furthermore, it has a tendency to bio-accumulate in the food chain. Generally, mercury is present in coal in concentrations well below 1 ppm; however, the large tonnages of coal consumed for electric power generation represent a significant source of mercury entering the environment. It is estimated that about 48 tons of mercury is accumulated in the environment annually in USA alone from coal-fired power plants.[1] Under Title III of the 1990 Clean Air Act Amendments, EPA is empowered to set emission standards for 189 pollutants including mercury. During coal combustion, the mercury is released into the exhaust gas as mercury vapor. This vapor may be oxidized to Hg^{2+} via homogeneous (gas-gas) or heterogeneous (gas-solid and surface catalyzed) reactions. Elemental mercury, once emitted, generally remains aloft in the atmosphere for much longer period and over greater distances and tends to travel regional or global distances before depositing to the earth's surface.

There are various technologies available to control the emission of mercury from coal-fired power plants. Among them is activated carbon injection into the flue gas stream. This technology has been studied for many years and is considered to be the "standard technology" at this time to control the mercury emissions from the flue gas. It can remove mercury by as much as 90% if activated carbon at a level of 4000 times of its stoichiometric amount is injected. The requirement of this excessive amount of activated carbon may be due to the fact that the adsorption of mercury species on activated carbon is not thermodynamically very favorable at the high flue gas temperatures (i.e., 150°C), and thus only the activated carbon that has high adsorption capability can be involved in the removal process. This makes the technology very expensive. For example, it is reported that the cost of mercury control is estimated to range from \$33,000/lb mercury removed to \$131,000/lb mercury removed for the bituminous coal-fired power plant unit.[2]

Theory

The removal of mercury from coal is based on the chemistry in which mercury is oxidized by a strong oxidizing reagent which is a combination of SO_2 and oxygen which are both self-supplied from the flue gas. SO_2 is a reducing reagent, but when it combines with oxygen in an acid solution, it becomes a strong oxidizing reagent. The cathode reaction for the oxidizing reagent can be written as:

$$H_2SO_3 + O_2 + 2e = SO_4^{2-} + H_2O \quad E^o = 2.29 \text{ eV} \tag{1}$$

where H_2SO_3 represents the aqueous SO_2. It can be seen that the magnitude of E^o value of reaction 1 is very high. This value is much higher than those of strong oxidizing reagents such as permanganate (1.51 eV) and dichromate (1.33 eV).

When SO_2 is dissolved in an aqueous medium, it forms H_2SO_3 in an acid solution which is then dissociated into HSO_3^- as pH is increased as shown:

$$SO_2 + H_2O = H_2SO_3 \quad K = 1.204 \tag{2}$$
$$H_2SO_3 = H^+ + HSO_3^- \quad K = 10^{-1.88} \tag{3}$$

The equilibrium constants of reactions 2 and 3 are those at room temperature. In an acid solution, dissolved SO_2 forms H_2SO_3. When the pH is increased, H_2SO_3 turns to HSO_3^-. The concentrations of H_2SO_3 and HSO_3^- are the same at pH 1.88. Thus, in order to increase the H_2SO_3 concentration in the dissolved SO_2, the solution pH should be below 1.88.

The dissolution of coal pyrite in the presence of SO_2 and oxygen may be based on an electrochemical reaction. The anode reaction may be given as

$$FeS_2 + 8H_2O = Fe^{2+} + 2SO_4^{2-} + 16H^+ + 14e \quad E^o \text{ (for cathode reaction)} = 0.355 \text{ eV} \tag{4}$$

The cathode reaction is reaction 1. The difference in E^o between cathode and anode reactions is large, suggesting that the leaching reaction is thermodynamically very much spontaneous under normal conditions. The overall leaching reaction can then be written as

$$FeS_2 + H_2O + 7H_2SO_3 + 7O_2 = Fe^{2+} + 9SO_4^{2-} + 16H^+ \tag{5}$$

The mercury sulfide upon liberation may be leached based on an electrochemical reaction. The anodic reaction may be written as:

$$HgS + 4H_2O = Hg^{2+} + SO_4^{2-} + 8H^+ + 8e \quad E^o \text{ (for cathode reaction)} = 0.54 \text{ eV} \tag{6}$$

It is seen that the difference in E^o between cathode and anode reactions is large, suggesting that the leaching of HgS is thermodynamically spontaneous under normal conditions. In the case of organic mercury, it may be seen as unoxidized or elemental mercury. The anodic reaction of elemental mercury is given as

$$Hg = Hg^{2+} + 2e \qquad E^{o} \text{ (cathode reaction)} = 0.85 \text{ eV} \qquad (7)$$

It can be seen from reaction 7 that the magnitude of E^{o} is high enough to be comparable with that of a precious metal such as silver (0.8 eV).

Experimental Methods

Sample Preparation

Coal samples of Illinois No. 6 coal were obtained from Peabody Energy. The coal was crushed and screened to give a fraction of 35 - 65 mesh which was used in this study. This size fraction of coal sample was analyzed for ash, nonpyritic iron, pyritic sulfur, and total sulfur using ASTM methods. The results showed that the non-pyritic iron was 0.13%, pyritic sulfur was 1.12%, total sulfur was 2.71%, and the ash content was 15.8%. This size fraction was submitted to a commercial analytical laboratory for mercury analysis. The mercury content was found to be 0.216 ppm. The raw coal was used because cleaning with froth flotation would decrease the pyrite content, and would result in difficulty in particularly analyzing for leaching of smaller amount of pyrite.

Leaching Experiments

The apparatus consisted of three gas cylinders (nitrogen, oxygen and SO_2) and a one-liter reactor as shown in the Figure 1. The reactor was immersed into a thermostated constant-temperature bath. It has four necks: the central neck is equipped with an overhead stirrer connected to a motor operating at a speed of 500 rpm. One of the side necks was fitted with a condenser, another with a bubbling dispersion tube, one with stirrer and the last with a sample-taking device. The gases, after metered for their gas flow rates, were combined and introduced into the reactor and bubbled through the coal slurry added to the reactor. During the leaching, the SO_2 and oxygen were dissolved and oxidized mercury content and coal pyrite. The undissolved fraction of SO_2 exited the reactor, and flowed through the condenser to knock out its moisture content. This gas stream was emitted into the atmosphere under a ventilation hood.

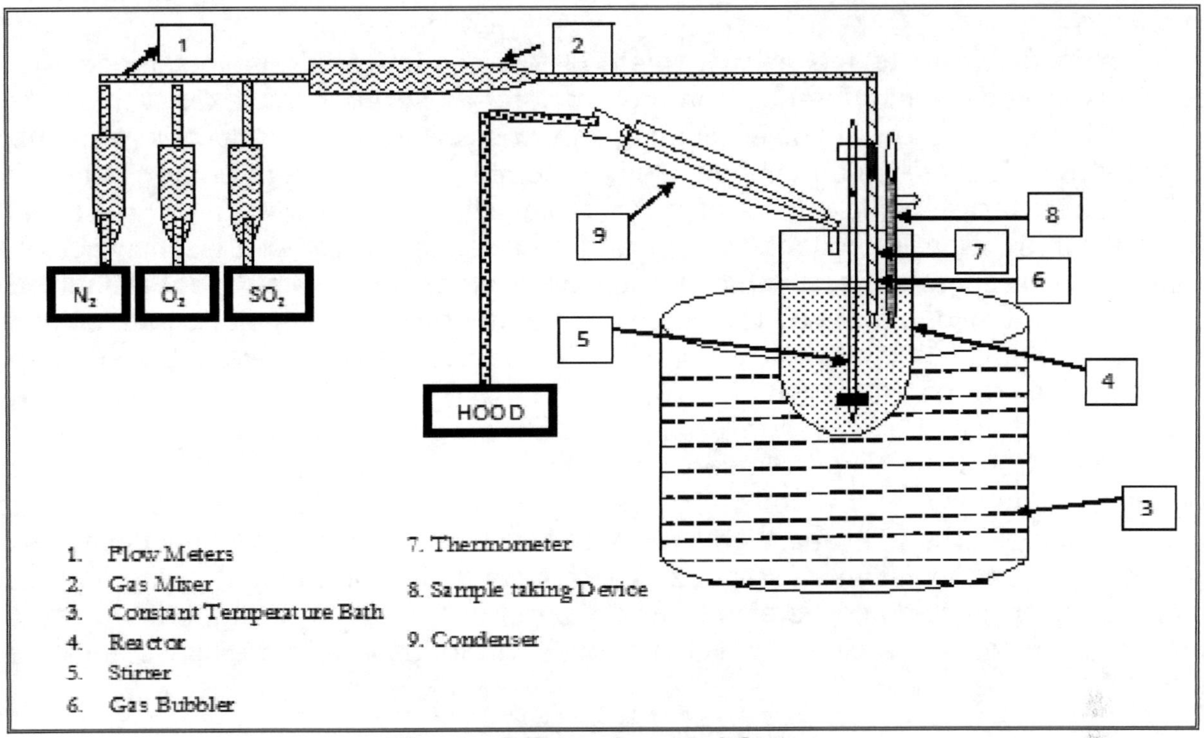

Figure 1 Experimental Setup

Leaching experiments were conducted by adding 50-g coal into 500 ml deionized water in the reactor and by stirring the mixture continuously with the overhead stirrer. The experiments were carried out in order to determine the effects of various parametric conditions on the mercury removal. The temperature was varied from 50 to 80°C. The solution pHs were 5.7, 1.82, 1.52, 1.25 and 1.08. The pH 5.7 was of the natural pH and the rest were achieved by adding 17.4, 35, 70 and 100 ml of 1 N sulfuric acid, respectively. The reaction time was 2 hours for all the experiments. The gas flow rate was maintained at 30 ml/s for all the experiments which contained 1000 ppm SO_2, 10% O_2 and the rest being N_2.

Liquid samples of about 20 ml were withdrawn from the reactor at 0.5, 1 and 2 hours, filtered and analyzed for soluble iron and mercuric ion. The analysis results were used to determine the removal percentages of coal pyrite and mercury content. They were calculated by taking the leaching portions over the original quantity of pyrite or mercury.

Samples were analyzed by an atomic absorption unit. Mercuric ion was analyzed by modifying an existing method. The existing method was based on the reduction of mercuric ion to elemental mercury in the presence of sodium tetrahydroborate before being analyzed as the cold vapor in a mercury hydride system connected to an atomic absorption unit. The description of the method may be available in the manufacture's manual of atomic absorption spectrophotometer. The problem with this existing method was that it suffered from interference with soluble iron. During the reduction of mercuric ion, the soluble iron was also reduced to elemental iron and precipitated. Mercuric ion seems to be co-precipitated with the iron, which reduced the absorbance to little or no value. The soluble iron came from the dissolution of coal pyrite and non-pyritic iron.

The modification to the existing method for mercury analysis was made by adding citrate. This solved the interference from iron as it forms strong complexes with soluble iron[8] but not with mercuric ion; thus, eliminating the possibility of the reduction of soluble iron during analysis. The modified method called for combining 9 milliliters of the filtrate with 3 ml of a 7.5% nitric acid solution, 3 ml of a 0.4 molar citrate solution and 100 μL of a 1 ppm mercury stock solution, thoroughly mixing and then taking 10 ml for the analysis. Standard solutions were prepared according to the following procedures: Nine ml of deionized water was mixed with 3 ml of a 7.5% nitric acid solution and 3 ml of a 0.4 molar citrate solution after which 100, 200 and 300 μL of a 1 ppm mercury stock solution were added to prepare standard solutions containing 6.62, 13.16 and 19.61 ppb mercury, respectively. The maximum absorbance value was taken from a data acquisition module-computer system for each analysis run. The concentration of mercury was determined by extrapolating the absorbance value into a calibration curve of absorbance versus concentration for the standard solutions. The plot was slightly deviated from the linearity probably because 3 ml of the 0.4 molar citrate solution was added. The concentration of mercury in the liquid sample was converted to leaching conversion (fraction of mercury reacted) using the original mercury content (0.216 ppm) as obtained from Standard Laboratories Inc., in the coal sample used.

Results

Figure 2 shows the effect of temperature on the mercury removal at the initial pH 5.7. The temperature was varied from 50 to 80°C. The final pH reached around 1.3 at 2 hours in the case of 71°C. It is seen that the removal of mercury increases as temperature increases. The mercury removal increases from 55.5% at 50°C to 99.2% at 75°C in two hours, but it slightly decreases to 94.7% at 80°C. The decrease in conversion when the temperature increased from 75 to 80°C may be due to the decrease in solubility of SO_2 and O_2.[9] This phenomenon may be related to the observation by Adams and Mattew [10] that the leaching rate of sphalerite (ZnS) with SO_2 and O_2 showed a maximum around 85°C.

Figure 3 shows the pyrite removal percentages at 71 and 75°C and pH 5.7 from the two same experiments as for Figure 2. The pyrite leaching conversion here and the values which will show up later in this paper are not exactly the values as the term implies. The leaching conversions were calculated over the combined amount of pyritic and non-pyritic irons. This was because the soluble iron in the solution could not be distinguished from the source of pyritic iron or non-pyritic iron. However, this calculation method and the method using the pyritic iron as the basis may not be much different because the non-pyritic iron content is only 11.6% of the pyritic iron. It is seen from Figure 2 that the pyrite conversions are very low. They are 1.2 % at 71°C and 1.8% at 75°C and after 2 hours. It is also clear from Figures 1 and 2 that the mercury removal percentages are much higher than those of pyrite. This result suggests that a substantial amount of mercury is not associated with coal pyrite. There must be then other forms of mercury which are co-existent such as HgO.

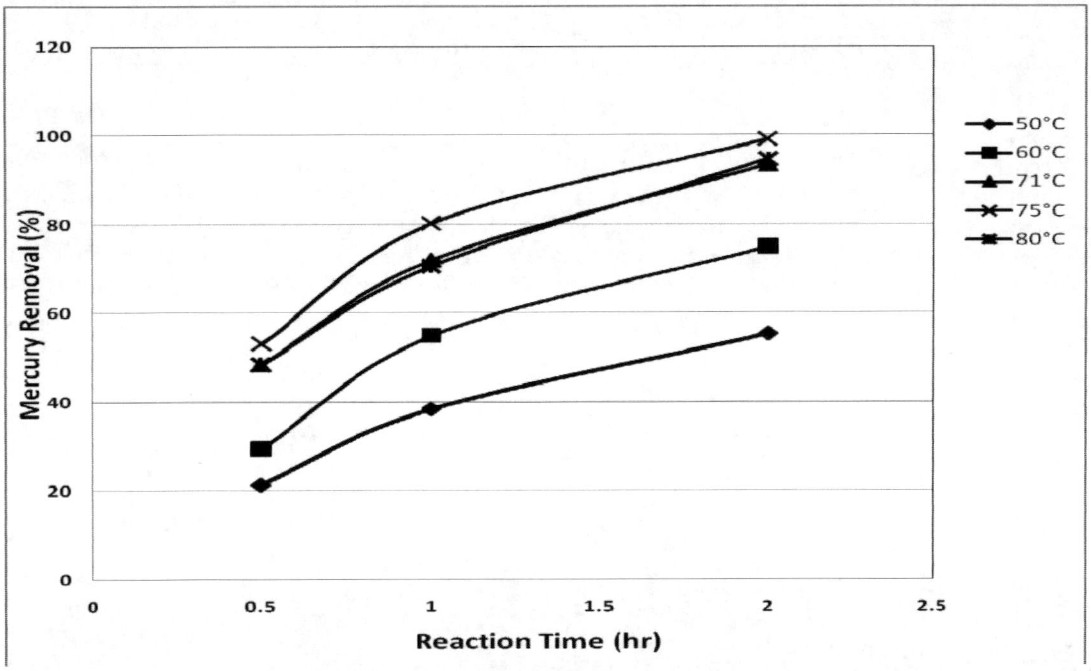

Figure 2 Percentage removals of Mercury at pH 5.7 and varying temperatures for Illinois No.6 Coal

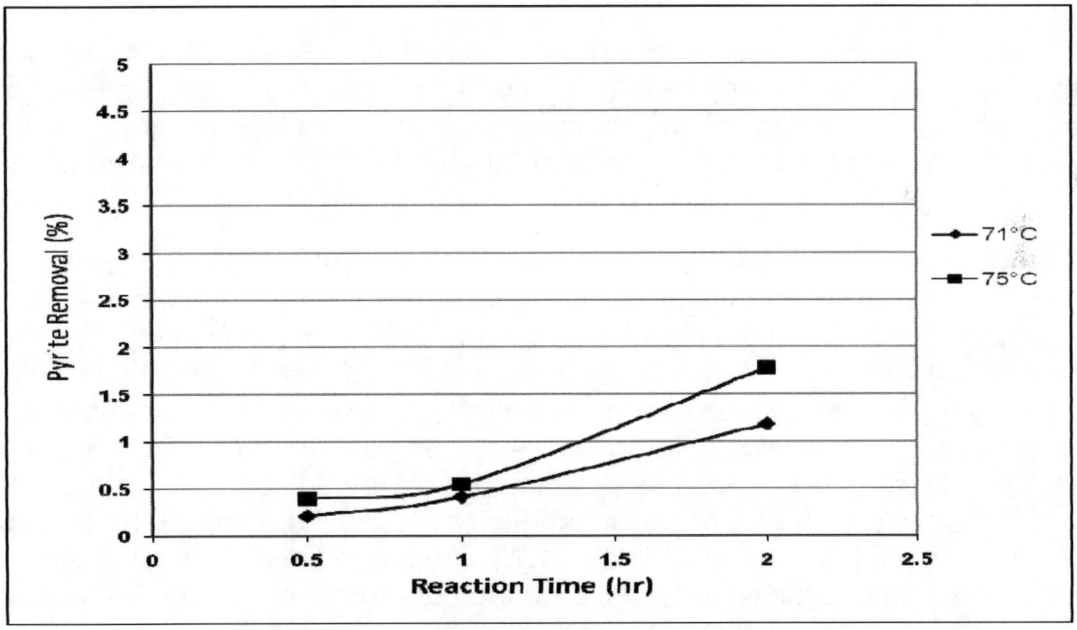

Figure 3 Percentage removals of Pyrite at pH 5.7 and varying temperatures for Illinois No.6 Coal

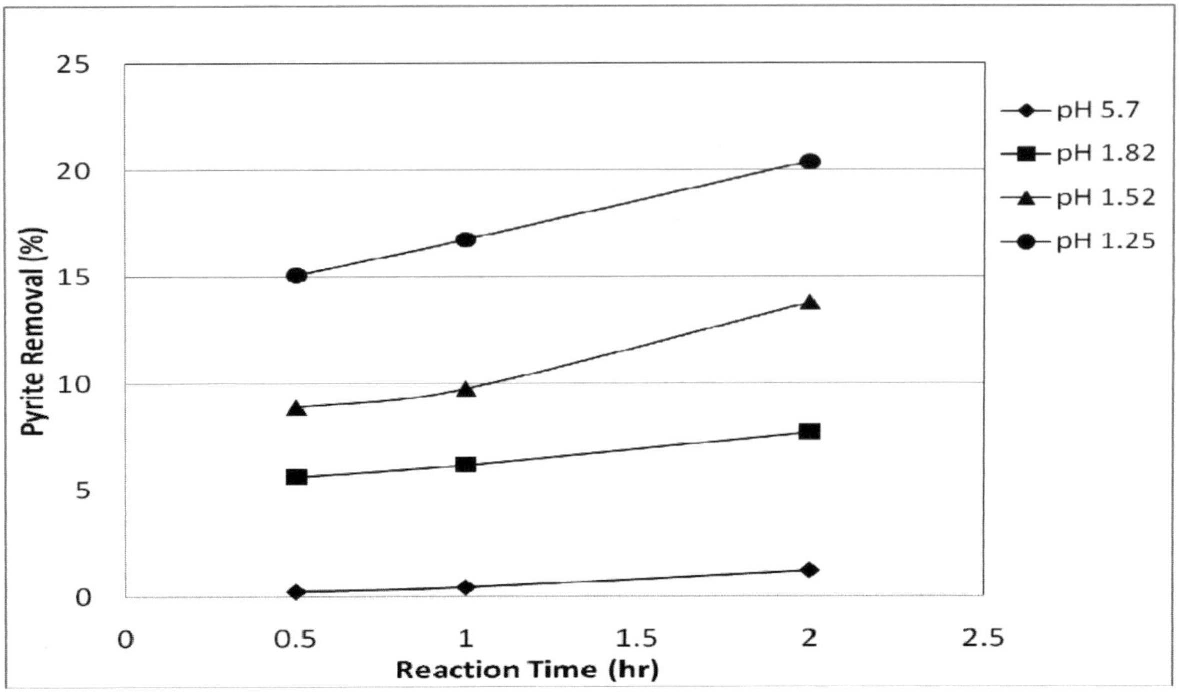

Figure 4 Percentage removals of Pyrite at 70.5° C and at varying pH values for Illinois No.6 Coal

Figure 4 demonstrates the effect of pH on pyrite removal at 70.5°C. The initial pH levels were 5.7, 1.82, 1.52 and 1.25. It is seen from above figure that the pyrite removal percentages increased with decreasing initial pH. It is seen that the effect is large as the removal percentage increases from 1.2 % at the initial pH 5.7 to 20.4% at the initial pH 1.25 after 2 hours.

Figure 5 demonstrates the effect of pH on mercury removal at 70.5°C. The initial pH levels were 5.7, 1.82, 1.52 and 1.25. As the leaching reactions of both pyrite and mercury (reactions 5 and 6) are acid producing, by the end of two hours the pH of the final solution in the reactor went down to 1.4, 1.28, 1.16 and 1.05 respectively. It is seen from above figure that the mercury removal percentages increased with decreasing initial pH. This may be due to the fact that more H_2SO_3 was available at the lower pH values as mentioned previously with reactions 2 and 3. Also, it is seen that the mercury removal is almost complete at the initial pH values of 1.52 and 1.25 at 2 hours. It can also be seen that mercury removal reached as much as 100% in two hours at initial pH of 1.25. This high value strongly suggests that this leaching scheme has a potential to be used for mercury removal in a commercial process.

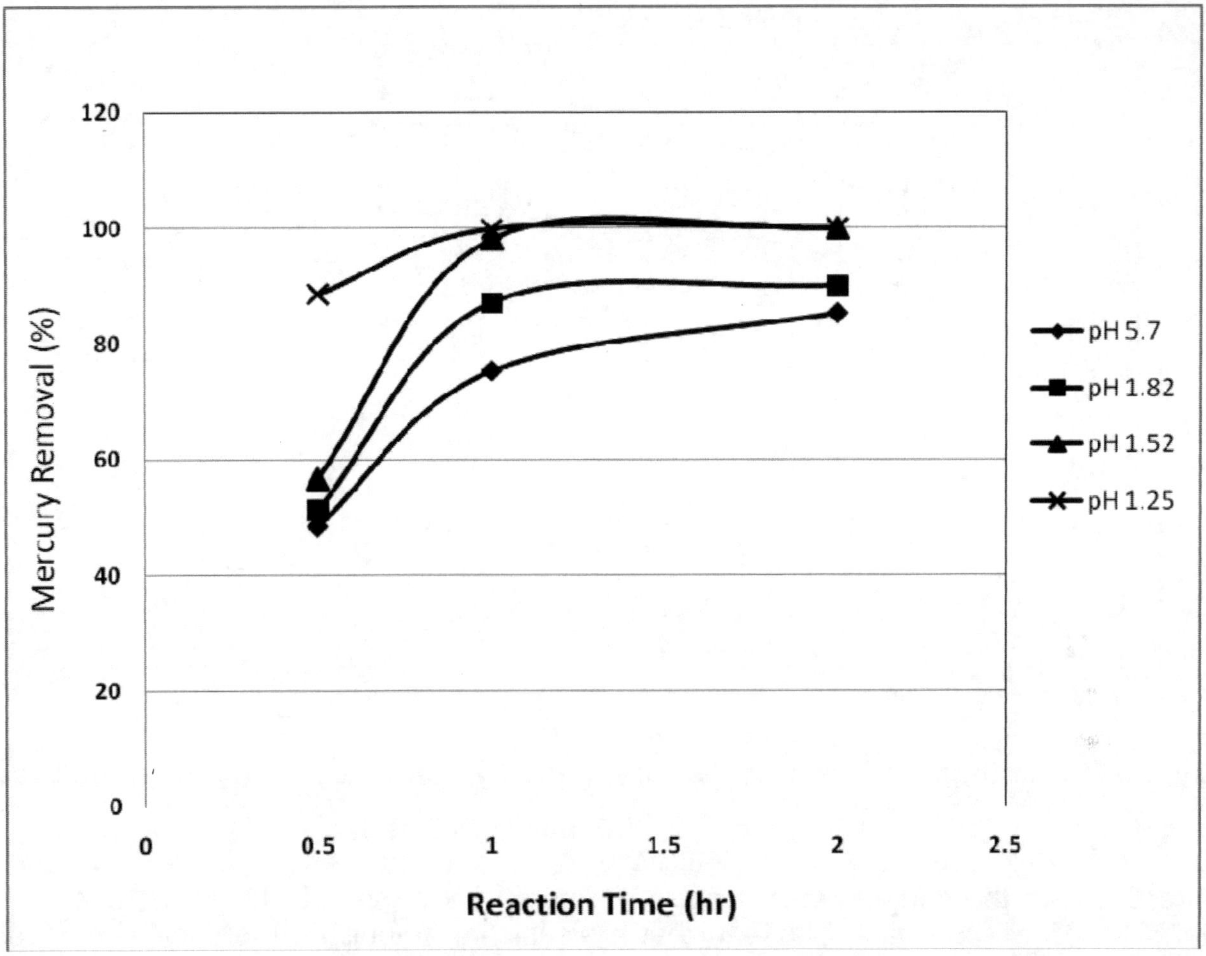

Figure 5 Percentage removals of Mercury at 70.5° C and at varying pH values for Illinois No.6 Coal

Acid Only Leaching (No Gas)

Two series of experiments were conducted to remove the mercury with only sulfuric acid: One is at the various temperatures and the other is at the various pH values. These series were initiated because it had been speculated that a significant fraction of the original mercury is non-pyrite association like HgO.[6]

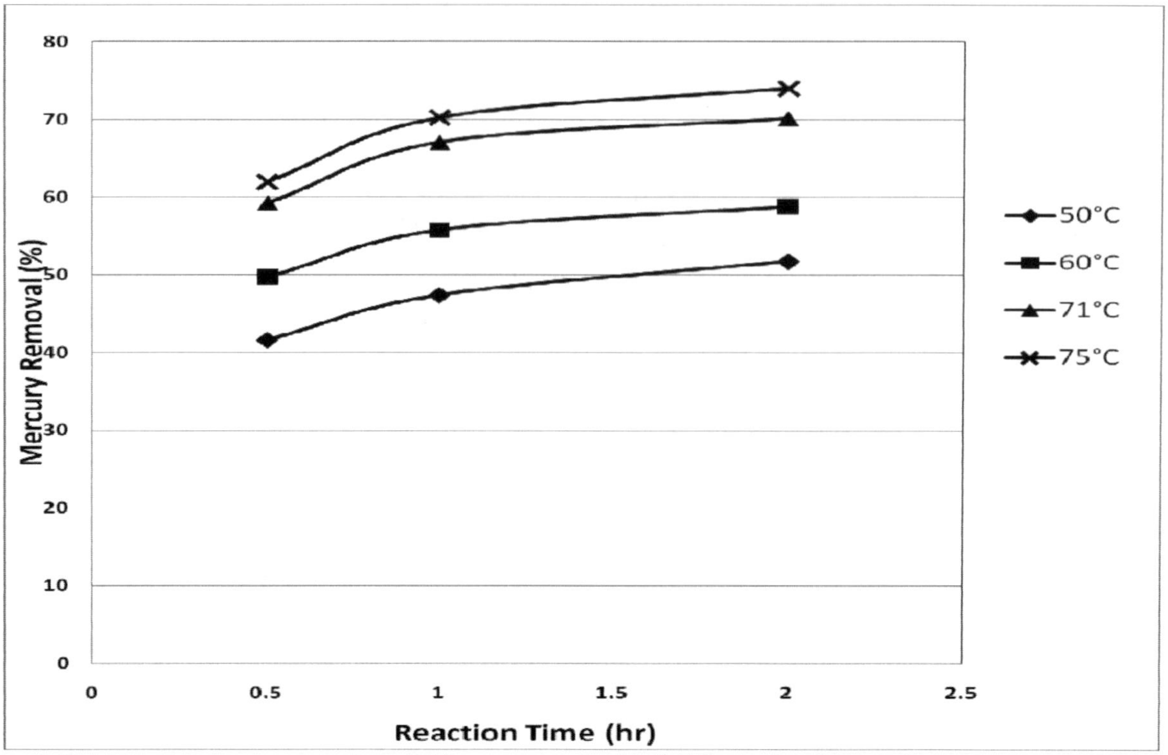

Figure 6 Percentage removals of Mercury at pH 1.25 and at varying temperatures and no gas for Illinois No.6 Coal

Figure 6 shows the effect of temperature on mercury removal at 50, 60, 71 and 75°C and initial pH 1.25 with no gas. It is seen that a significant amount of mercury is removed with sulfuric acid alone without any gas. It reaches 74% at 75°C after 2 hours of reaction time. From this it can be said much of the mercury in Illinois No.6 coal may be acid soluble HgO. However the fraction of HgO in the Pittsburgh No.8 coal seems to be much lower as the leaching reaction under same conditions yielded only 40% mercury removal[12]. It can be suspected that the gas (SO_2) acts as a medium used to increase the concentration of H_2SO_3 so as to facilitate the leaching reaction and does not necessarily take part in the leaching reaction.

Figure 7 shows the effect of initial pH on mercury removal at 70.5°C and no gas. The initial pH was varied from 1.82 to 1.08. Again, it is seen that a significant fraction of mercury is removed. It reaches 73% at the initial pH 1.08 and 71°C. The mercury removal percentages at pH 5.7 were little or none. It is obvious that approximately 75% of the original mercury is acid soluble. Much of it could be HgO.

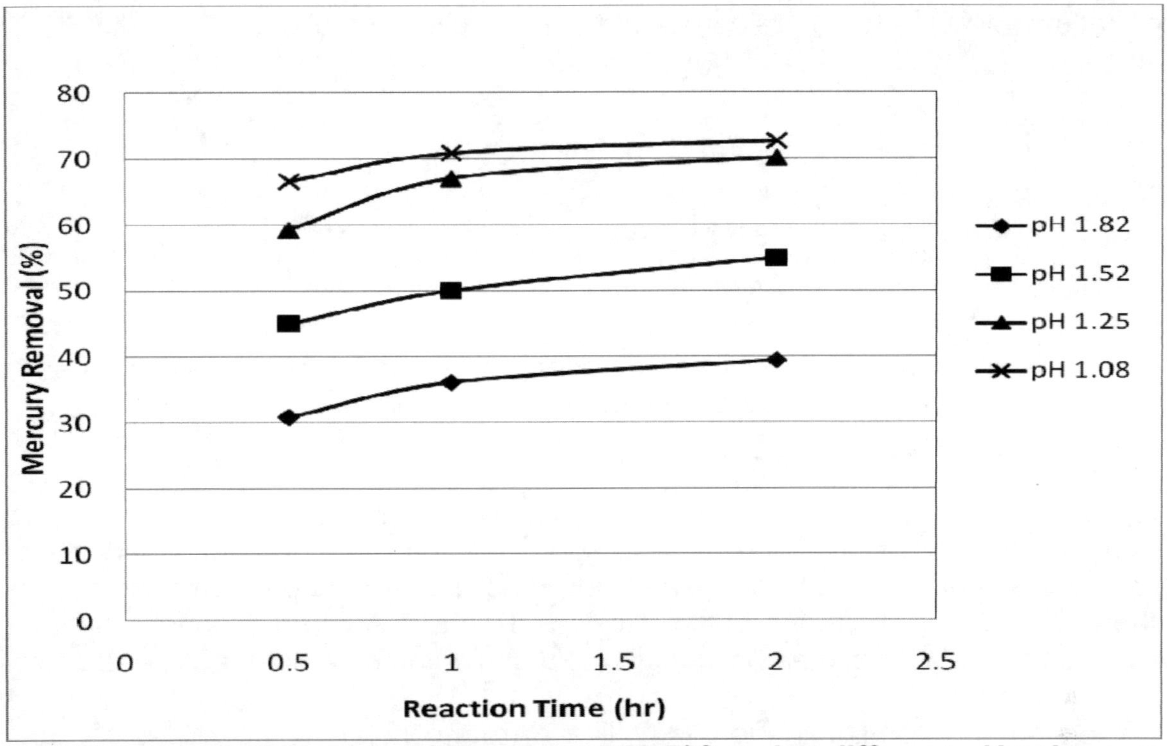

Figure 7 Percentage removals of Mercury at 70.5°C and at different pH values and no gas for Illinois No.6 Coal

Discussion

In this study, the leaching rate of mercury was analyzed by the reaction zone model. This model is based on a shrinking core model in which reaction zone in the reaction interface moves inward.[5, 11] The reaction zone may be viewed as a region in which disseminated mercury grains are in all stages of reaction. Here the mercury grains represent grains of elemental mercury, organic mercury, HgS and HgO. As the zone moves to smaller values of r (radius of unreacted core), unreacted mercury grains are included in the zone. At the outer edge of the zone, the last vestige of unreacted mercury disappears. The leaching rate based on chemical reactions within the zone may be given as

$$\frac{dn}{dt} = -\frac{4p\, r^2}{f} d\, n_g A_g [SO_2]^p [O_2]^q k \qquad (8)$$

where n = number of moles of unleached mercury in a coal particle at time t, δ = reaction zone thickness, n_g = number of mercury grains per unit volume of coal in the reaction zone, A_g = average surface area of a mercury grain, $[SO_2]$ and $[O_2]$ concentrations of SO_2 and O_2 in the reaction zone, respectively, p = reaction order with respect to $[SO_2]$, q = reaction order with respect to $[O_2]$, k = rate constant, F = sphericity of coal particle.

After integrating and simplifying this equation under the assumption that the concentrations of SO_2 and O_2 are constant during reaction, we obtain

$$1 - (1 - F)^{1/3} = kt \qquad (9)$$

$$where\ k = \frac{3Md}{f\, r_g r_o r_g}[SO_2]^p[O_2]^q k \qquad (10)$$

where M = average molecular weight of mercury grains, r_g = average mercury grain radius, r_o = coal particle radius, ρ_g = average density of mercury grains, and F = fraction of mercury reacted, or conversion.

The leaching data for mercury which are given in Figure 2 are plotted according to Equation (9), and shown in Figure 8. It is seen that the plots produce good straight lines. However, the straight lines do not pass through the origin. This phenomenon was observed with the same type of plots for Pittsburgh No. 8 coal.[12] Again, this phenomenon may be evidence that there is more than one single reaction having different leaching rates.

The slope of each line in Figure 8 will be k (rate constant). The values of rate constant K are tabulated in Table 1.

Table 1 Rate Constant Values for Mercury Leaching

Serial No.	Temperature (°C)	Rate Constant (k_1) (1/hr)
1.	50	10.3×10^{-2}
2.	60	16.8×10^{-2}
3.	70.5	26.2×10^{-2}
4.	75	38.1×10^{-2}

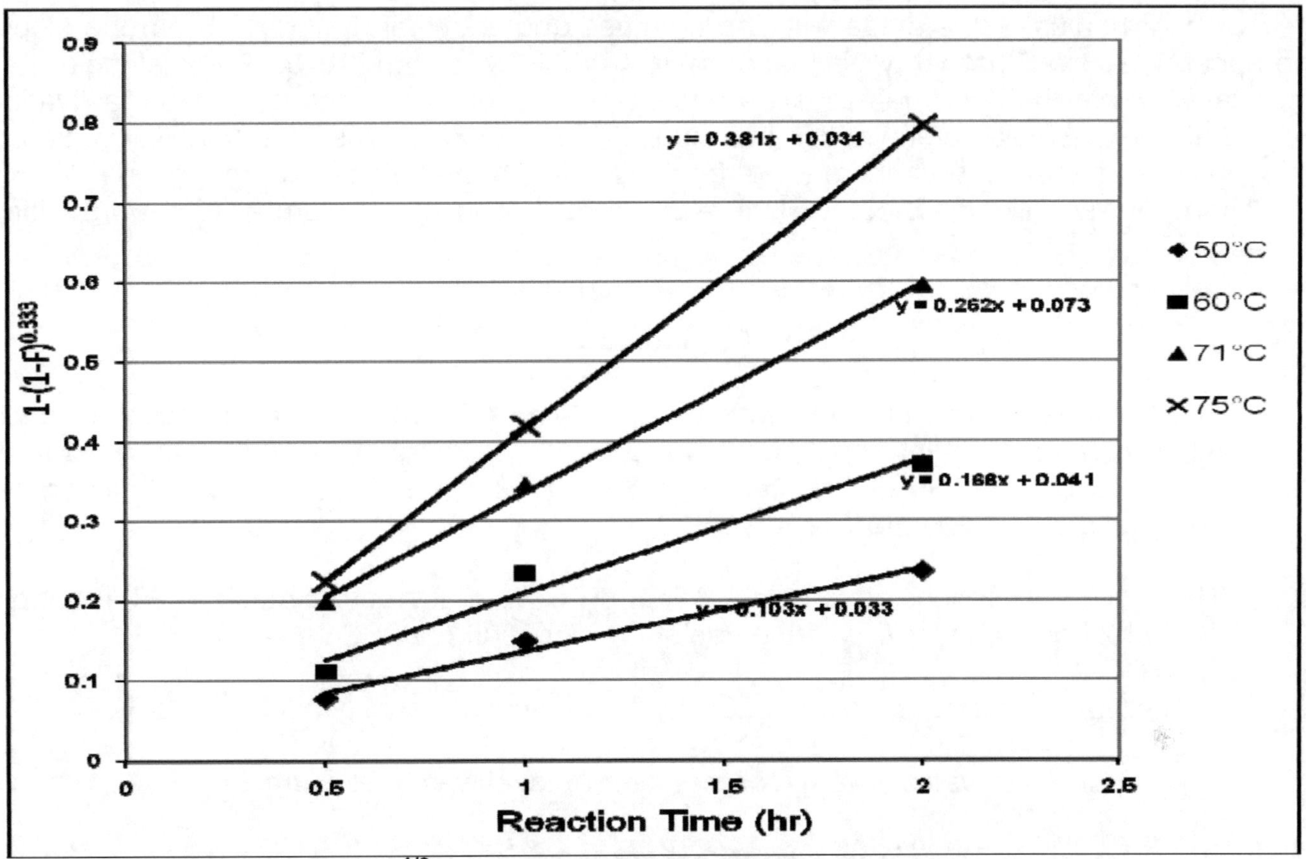

Figure 8 Plot of [1-(1-F) $^{1/3}$] vs time at various temperatures for mercury removal for Illinoi No.6 Coal

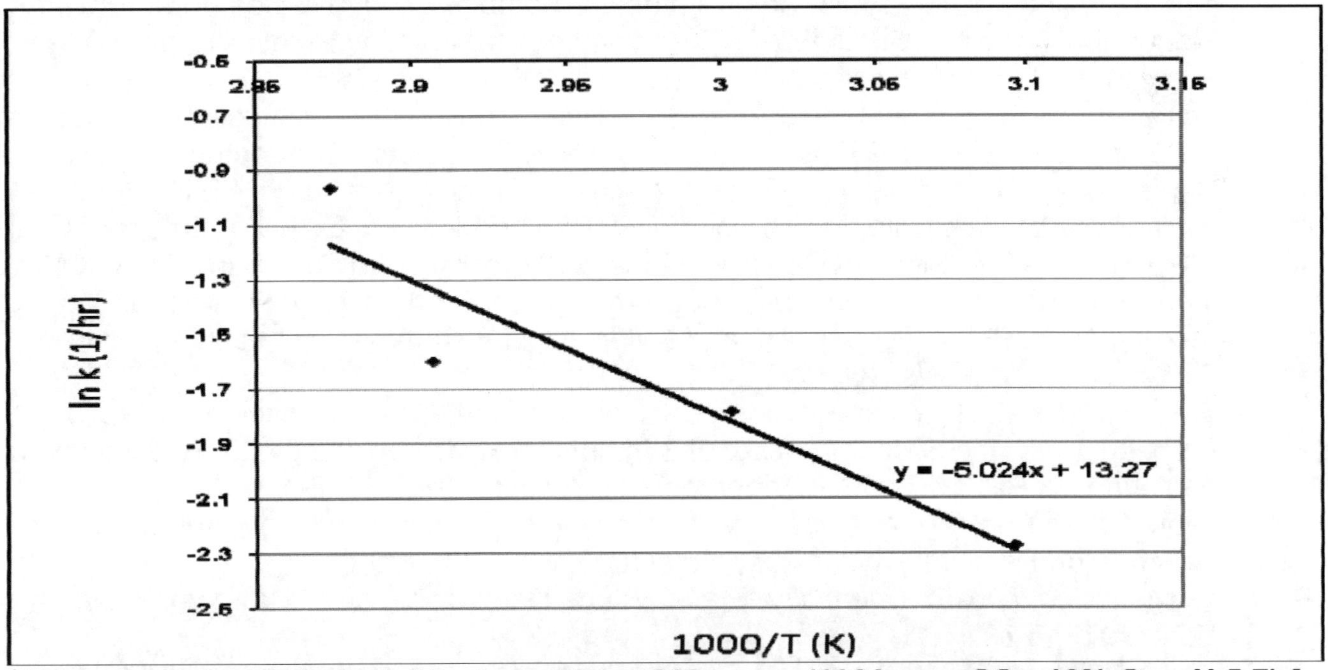

Figure 7.3.2 Arrhenius Plot for Leaching of Mercury (1000 ppm SO_2, 10% O_2, pH 5.7) for Illinois No.6 Coal

An Arrhenius plot is made with the k values and shown in Figure 9. From the slope, the apparent activation energy was calculated and this was found to be 9.9 kcal/mole. It is important to note that the true activation energy may be higher than the apparent activation energy of 9.9 kcal/mole determined in this study. This may be because the solubility values of SO_2 and O_2 decrease with increasing temperature.[9] Nonetheless the magnitude of this activation energy validates the chemical reaction controlled mechanism assumed in writing Equation (8).

Conclusions

1. The oxidizing power of combination of SO_2 and oxygen is strong enough to leach much mercury content from Illinois No. 6 coal. The mercury removal increased from 55.5% at 50°C to 99.2% at 75°C, pH 5.7 and 2 hours of reaction time under the conditions of 1000 ppm SO_2 and 10% O_2.

2. The pyrite conversions were only 1.2% at pH 5.7 and increased much to 20.4% at pH 1.25 at 2 hours and 71°C with the gas stream of 2000 ppm SO_2 and 10% O_2.

3. A significant amount as much as 75% of the original mercury is acid soluble.

4. The apparent activation energy for mercury removal was 9.9 kcal/mole.

Literature References Cited

1. http://www.epa.gov/camr/basic.htm
2. Hoffman, J.; and Brown, J. R. Preliminary Cost Estimate of Activated Carbon Injection for Controlling Mercury Emissions from an Un-Scrubbed 500 MW Coal-Fired Power Plant. Final Report to US DOE/NETL (Innovations for Existing Plants Programs) by Science Applications International Corporation. November **2003**.
3. Luttrell, G. H.; Yoon, R. H.; and Adel, G. T. Precombustion Removal of Hazardous Air Pollutant Precursors. **1997** Coal Liquefaction and Solid Fuels Contractors Review Conference – Hazardous Air Pollutant Precursor Control; Pittsburgh, FETC/DOE.
4. Rosandale, L. W.; Devito, M. S.; Conrad, V. B.; and Meenan, G. F. The Effect of Coal Cleaning on Trace Element Concentrations. **1993**, an internal document, CONSOL Inc., Research and Development, 4000 Brownsville Road, Library, PA 15129.
5. Sundaram, H. P.; and Cho, E. H.; and Miller, A.. SO_2 Removal by Leaching Coal Pyrite. *Energy & Fuels*, **2001**, 15(2); 470-476.
6. Cho, E. H.; Yang, Ray Mercury Removal from Coal by Leaching with SO_2. Proceedings of the Sixth International Symposium Hydrometallurgy **2008**, edited by C. A. Young, P. R. Taylor, C. G. Anderson and Y. Choi, SME, 1143-1153.
7. Jeon, C.; and Holl, W. H. Chemical Modification of Chitosan and Equilibrium Study for Mercury Ion Removal. *Water Research*, **2003**, 37; 4770-4780.
8. *Lange's Handbook of Chemistry*, Eleventh ed.; Dean J. A.., ed.; McGraw-Hill: New York, **1973**, 5-53.
9. *Lange's Handbook of Chemistry*, Eleventh ed.; Dean J. A.., ed.; McGraw-Hill: New York, **1973**, 10-7.
10. Adams, R. W.; and Mattew, I. G. Leaching of Sulfide Concentrates at Atmospheric

Pressure Using SO_2 and O_2 Mixture. *Proc. Australian Ins. Min. and Metall.* Dec. **1981**, 280, 41-52.
11. Cho, E. H.; Chang, K. H.; and Rollins, R. R. A Kinetic Study of Leaching of Coal Pyrite with Nitric Acid. *Met. Trans. B*. September **1983,** 14B, 317-324.
12. Cho, E. H.; Chateker, P.; Garlapalli, R.; and Yang, R. Y. K. Mercury Removal from Coal by leaching with SO_2. *Energy Fuels*, 2009, 23 (2), pp 774–778

Acknowledgments

I greatly express my gratitude to CAST (Center of Advanced Separation Technology) sponsored by NETL/DOE under Grant CR-19147-429231 which made this research project possible. I greatly acknowledge the enthusiastic supervision of Dr.Eung Cho during this research. Also, acknowledged is Mr. David P. Byer, Senior Manager, Quality Control and Contract Labs, Peabody Energy for donation of an Illinois No. 6 coal sample which was used in this study.

PACKED BED REACTOR FOR CONTINUOUS ENZYMATIC RING OPENING POLYMERIZATION OF LACTONES

Santiago Faucher*, Lovina Pereira, and Jordan Wosnick

Xerox Research Centre of Canada

Abstract: Enzymatic Ring-opening Polymerization of lactones permits the production of polylactones at atmospheric pressures, low temperatures, and without the use of toxic tin catalyst. Because of these attributes this polymerization process has often been regarded as a more environmentally friendly route to the production of polyesters as compared to classical condensation chemistry. Enzymatic polymerizations however suffer from low reaction rates and as a result require the use of high supported catalyst loadings (10 wt% vs. monomer) in batch processes. Further increases in the catalyst concentrations add cost to the process, reduce batch throughput, and further complicate the downstream catalyst-product separation step given the economic need to recover and re-use the catalyst. When we examine these challenges closely we note that the separation step itself markedly deteriorates the environmental benefits of enzymatic polymerization over condensation polymerization. Large volumes of solvent are required to dilute the viscous reaction product sufficiently to permit filtration for catalyst recovery. To overcome this challenge we examined a number of possible reactor technologies and specifically studied the use of packed-bed continuous reactors to replace the batch process. Monomer solutions are fed through the reactor to produce, at the bed's exit, a catalyst-free polymer product. In addition, because of the higher enzyme concentrations in the packed-bed, reaction rates observed with this reactor configuration are substantially higher than those observed in batch. This permits the high throughput production of polymers and intensified experimentation at the small scale. As a result of this successful technology the sustainability of the enzymatic polymerization process has been greatly improved allowing the production of a wide range of co-polymers with 0.5% of the catalyst and 20% of the solvent originally used in the batch process.

Keywords: Enzymatic, Ring, Opening, Polymerization, Continuous, Co-polymer, Packed-Bed, Lipase, Sustainability

1. INTRODUCTION

Ring opening of lactones via enzymatic polymerization has been an interesting approach to the synthesis of bio-degradable poly(lactones) useful in biomedical applications (*Varma 2005*). This process offers the advantage of mild reaction conditions, low temperatures and atmospheric pressures, and the elimination of toxic metal catalysts in contrast to conventional condensation polymerizations. The process itself can be carried out as a bulk polymerization where only enzyme and monomer are loaded into a batch reactor. These traits support the widely held view that the enzymatic polymerization process is environmentally friendly. However, while the synthetic step in the polymerization offers environmental advantages, the downstream separation processes typically applied to the product for catalyst removal and recovery do not.

The supported enzyme has a low activity and so large quantities of enzyme, 10 wt% relative to monomer, are typically used for the production of polymer. This catalyst must therefore be removed from the polymer product so that the product is of the desired purity. Economics also drive this need for separation as the catalyst is costly and must be recycled to drive down the product cost. Separation of the catalyst from the polymer is complicated by the high viscosity of the product and so solvents are used to dissolve the product prior to filtration. This need for solvent use and the associated detrimental waste created reduce the environmental and economic attractiveness of this technology.

From a process development perspective, the low activity of the enzyme drives the use of higher catalyst concentrations so that reactions are undertaken in days rather than weeks. These high loadings however make the resulting polymer un-attractive from an economic perspective. We realized that to resolve this challenge by the conventional wisdom of recover-recycle would necessitate the use of large solvent volumes and deteriorate the sustainability of this technology. We thus turned to examine other reactor configurations that would allow high concentrations of catalyst per volume and avoid the need for polymer-catalyst separation post polymerization. We report herein the development of a continuous enzymatic polymerization process utilizing a packed bed reactor. This approach eliminates the need for catalyst recovery and recycling, permits the use of high catalyst/monomer ratios to accelerate polymerization rates, and greatly reduces solvent usage.

2. EXPERIMENTAL

2.1 Materials

Novozym®435 (resin immobilized Candida antarctica lipase B, Novozymes, Denmark) was dried under vacuum overnight. 11/12-Pentadecen-15-olide (Pentadecene-lactone, PDCE-L) and pentadecalactone (PDCA-L) were purchased from Ivanhoe and used as received. Toluene was used as received. Caprolactone (CP-L) and Glass beads (0.2 mm) were purchased from Sigma-Aldrich and used as received.

2.2 Typical Column Packing and Reactor Set-up

A 10-cm long segment of ¼-inch stainless steel tubing was capped at one end and filled with 70 volume % of 200-μm glass beads and 30 volume % (0.1 g) of Novozym® 435 to form the packed bed reactor. The contents of the packed bed reactor were equilibrated in toluene overnight to allow the resin to swell. After equilibration, an inlet feed line and outlet product line was attached to the packed bed reactor. The inlet feed-line was connected to a pump and the later to a feed tank containing the monomer feed. The outlet line was connected to a receiving tank. This arrangement is shown in Figure 1.

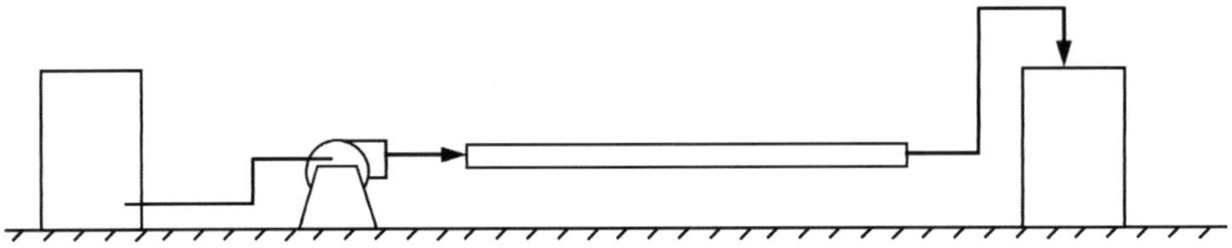

Figure 1. Process flow diagram for Continuous Enzymatic Polymerization of Lactones in a Packed-bed Reactor.

2.3 Typical Production of Polyesters

The packed bed reactor was placed in a heating bath preset to the desired temperature. A monomer solution containing one or more of PDE-L, PDA-L, CP-L and toluene was pumped through the packed bed reactor continuously at the desired flowrate. The product exiting the packed bed reactor was collected once the process became stable. Reactor process conditions were then modified and a second experiment undertaken subsequently. Polymer molecular weight and molecular weight distribution as well as monomer conversion were measured utilizing gel permeation chromatography (GPC).

3. RESULTS AND DISCUSSION

Our first continuous packed bed design relied on a simple 10 cm long stainless steel tube with ¼" OD within which we would load the enzyme along with glass beads. The function of the beads were two fold, firstly to dilute the active enzyme concentration so that reaction rates were sufficiently low to prevent the production of high molecular weight polymer (non-dissolvable in toluene) that we found would cause high back-pressures and block flow in the column, and secondly to prevent the extrusion of the supported enzyme catalyst we were using (Novozyme 435). The enzyme in Novozyme 435 is supported on a lightly cross-linked and solvent-monomer swellable support that deforms sufficiently under pressure to be extruded out of the bed along with the product under highly viscous conditions. By utilizing non-deformable glass beads, we felt that the large deformable catalyst support would be secured in the column more effectively and not elute with the product. In general this strategy proved successful and so we conducted a large number of preliminary experiments that showed a twenty fold increase in the productivity of the catalyst (mass of polymer produced/ mass of catalyst used) as compared to the traditional batch process. This beneficial increase in productivity came with a five fold reduction in the amount of solvent used. The only solvent used in this process was that necessary to maintain sufficiently low viscosities for pumping the reaction media-product through the packed bed.

The true productivity of the process was not however understood and so we continued the work with the above configuration to understand the retention of enzymatic activity with time. The twenty fold improvement had been derived from operating a column over an 8 hour period. Longer production times, where the monomer solution is continuously flowed through the same catalyst bed, would therefore likely yield higher productivities than the 8 hour production run.

Extended continuous production of co-Polymers of caprolactone and pentadecalactone

Figure 2 summarizes the set-up and results from an extended continuous production of co-polymers of caprolactone and pentadecalactone over 100 hours. Figures 3, 4 and 5 summarize the monomer conversions, polymer composition, and product polymer characteristics (M_w, M_n, PDI) as a function of time respectively. As can be noted in Figure 3 overall monomer conversions dropped with time from ~70% to ~20% over a period of 100 hours of operation. Interestingly the composition of the co-polymer also shifted with time with a reduced incorporation of caprolactone relative to pentadecalactone as enzyme activities decreased. Likewise polymer molecular weights were found to decrease with time, a sign of deteriorating enzyme activity. Polymer molecular weights ranging from 18,000 Da to 8,000 Da were produced during the experiment. From this experiment approximately 2000g of polymer was produced per gram of supported catalyst used. This showed an order of magnitude increase over our prior continuous work and making the productivities of this process 200 fold that of the batch process.

The mechanism by which the enzyme activity deteriorates with time is not clear to us. It is possible that the enzyme leaches out of the bed with time or that the enzyme denatures due to the high reaction temperatures and organic solvents used. Impurities in the system may also be at fault for denaturing the enzyme. Ideally if the enzyme activity could be retained then much higher productivities would be achievable. We have not yet studied this challenge however. It was more interesting from a sustainability (environmental) perspective to first resolve a second challenge with this process, the elimination of solvent from the process altogether to undertake continuous bulk polymerizations.

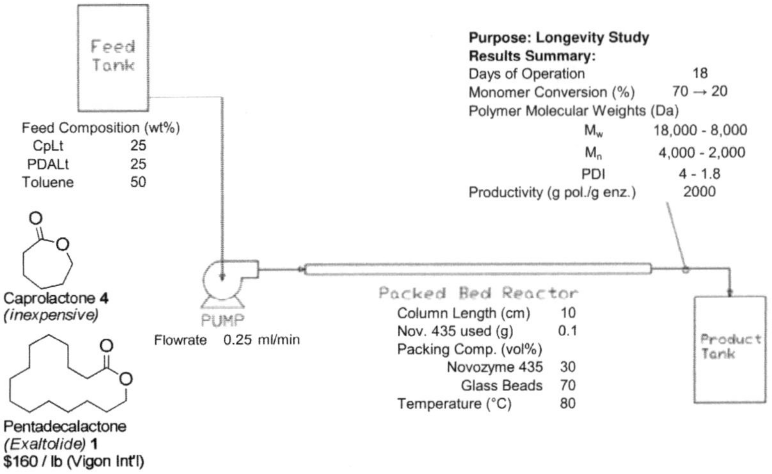

Figure 2. Set-up and summary of results for continuous enzymatic polymerization of caprolactone and pentadecalactone.

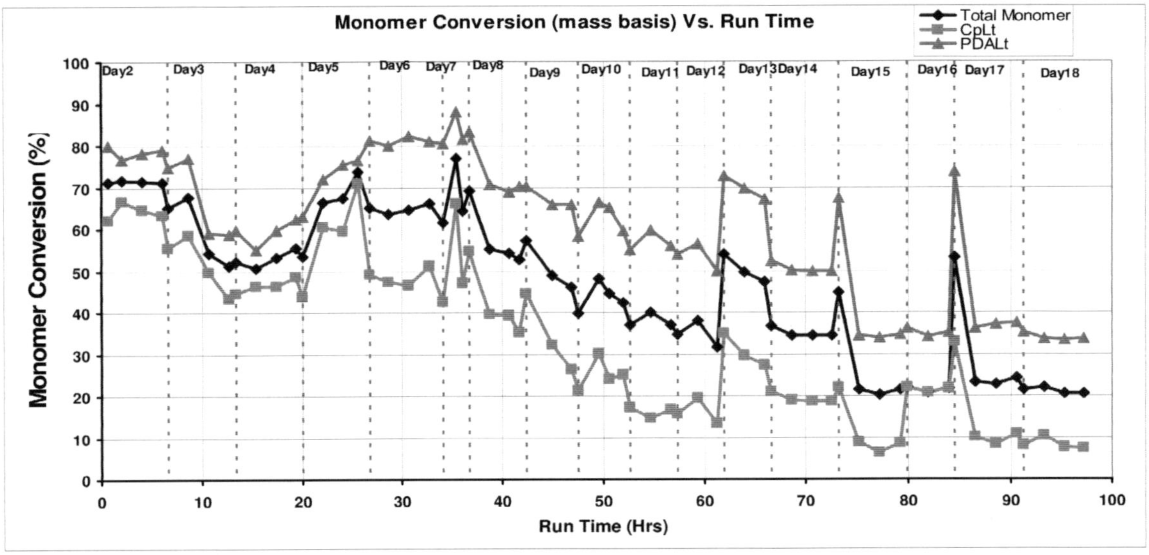

Figure 3. Monomer conversion versus operating time for the experiment described in Figure 2.

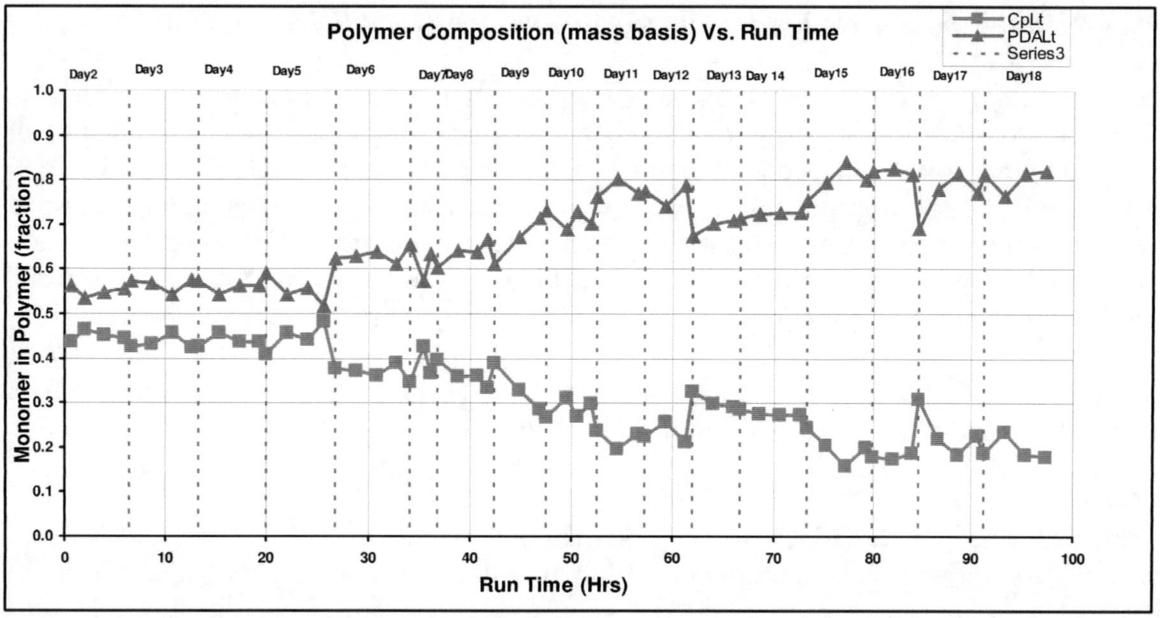

Figure 4. Polymer composition as a function of operating time for the experiment described in Figure2.

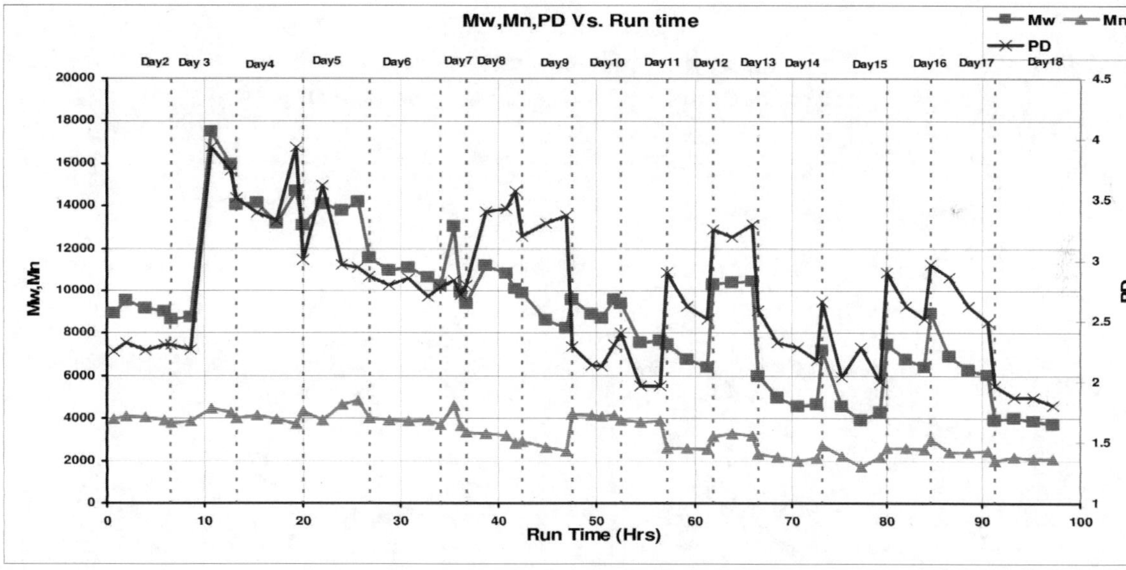

Figure 5. Polymer characteristics as a function of operating time for the experiment described in Figure 2.

Packing re-Design for Solvent Free (Bulk) Continuous Enzymatic Polymerization

While the extended experimental work was successful in demonstrating high catalyst productivities (two orders of magnitude higher than those obtained in batch) and the solvent usage in this process was reduced by a factor of 5 as compared to the batch process, the complete elimination of solvent was desirable from both an environmental and economic perspective. However bulk polymerizations using the glass-enzyme packing proved unsuccessful on account of the high back-pressure generated by this packing. Figure 6 shows the pressure drop as a function of face velocity for the glass-enzyme and enzyme packing. It seemed apparent from this characterization that to achieve lower pressure drops and permit bulk polymerizations we would need to replace the glass beads with a material that was closer to the porosity of the Novozyme 435 supporting material. A polystyrene-divinylbenzene bead was chosen as a substitute to the glass beads and experiments run in the absence of solvent as summarized in Figure 7.

With this new packed-bed, bulk polymerizations of 11/12-pentadecen-15-olide and pentadecalactone were achieved producing polymer with molecular weights ranging from 10,000 to 52,000 Da (M_w). Monomer conversions were however low (22%) indicating that a longer column or even lower monomer feed rates would be necessary to achieve complete conversions. We continue to study this approach to further increase conversions and attain ideally a continuous bulk polymerization system.

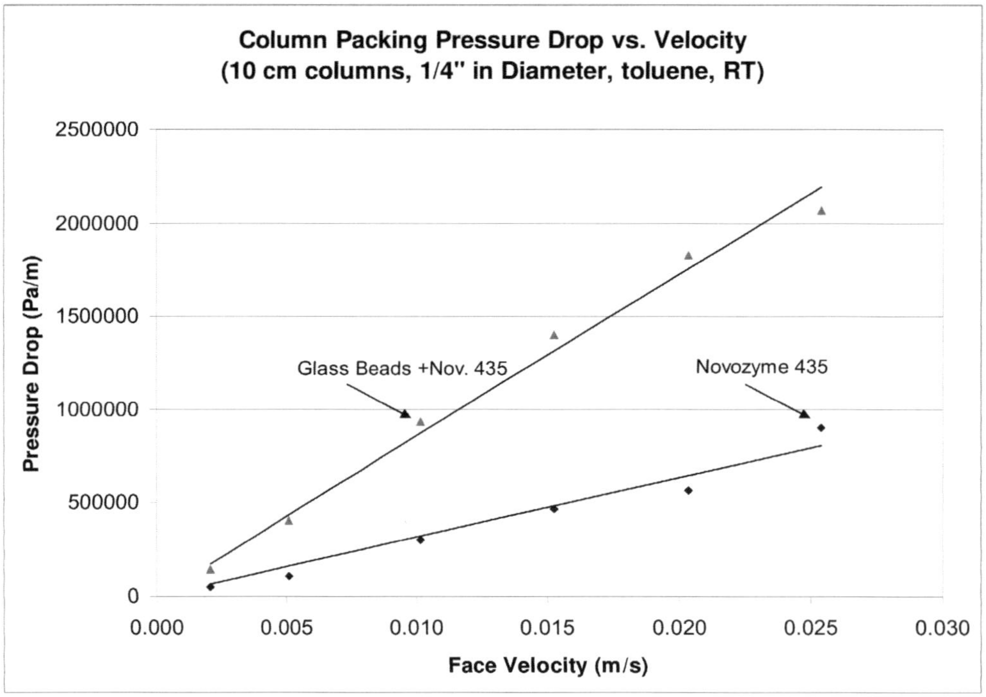

Figure 6. Back-pressure as a function of face velocity for packed bed consisting of glass bead + supported enzyme and for supported enzyme alone.

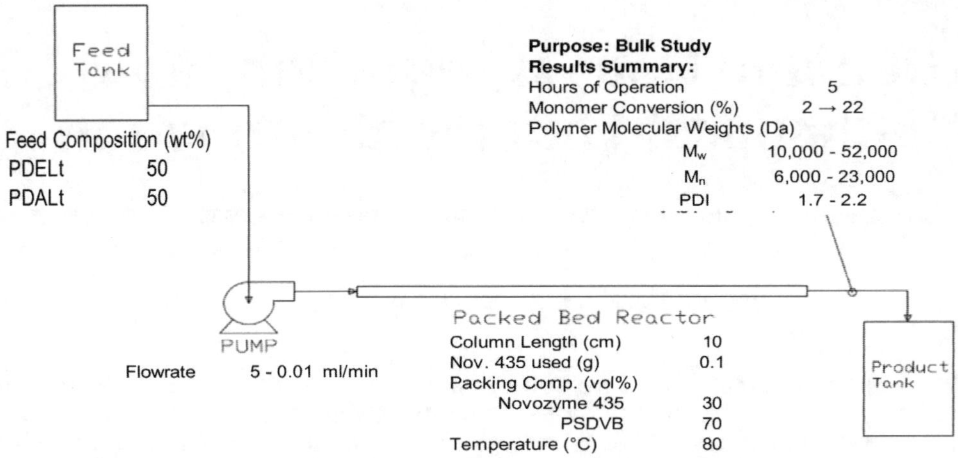

Figure 7. Set-up and summary of results for continuous bulk enzymatic polymerization of pentadecenelactone and pentadecalactone

4. CONCLUSIONS

Utilizing a continuous packed bed reactor, catalyst productivities that are two orders of magnitude higher than those achievable in a single batch experiment have been demonstrated. This process furthermore reduces solvent usage by a factor of 5 as compared to the current batch enzymatic polymerization technology. While these are significant improvements over the current state of the art we continue to develop solvent-free processes that will completely eliminate the use of solvent in enzymatic polymerization and make it a true green chemistry. We have demonstrated here that by careful selection of the catalyst packing bulk polymerizations are possible but that further process development work is necessary to achieve complete conversions.

REFERENCES

Varma, I.K., Albertsson,A.-C., Rajkhowaa, R., Srivastavaet R.K. (2005) Enzyme Catalyzed Synthesis of Polyesters, Prog. Polym. Sci., 30, 949–981.

Wosnick, J., Faucher, S. (2008) Production of Polyesters in a Continuous Packed-Bed Reactor using Immobilized Enzyme Catalysts, US Patent and Trademark Office Patent Application.

Optimization and Heat Integration of An IGCC Plant with Chemical Looping Combustion

Juan M. Salazar, Vishwamitra Research Institute, Center for Uncertain Systems: Tools for Optimization and Management, Clarendon Hills, IL

Urmila Diwekar, Vishwamitra Research Institute, Center for Uncertain Systems: Tools for Optimization and Management, Clarendon Hills, IL

Integrated Gasification Combined Cycle (IGCC) system using chemical looping combustion is an important approach for future energy options. This work focuses on the multiobjective optimization and water gas shift (WGS) reaction heat integration using ASPEN Plus, ASPEN energy analyzer and CAPE-OPEN compliant capabilities for multiobjective optimization and MINLP process synthesis developed by Vishwamitra Research Institute. The feasible operating surface for the IGCC system is generated and deterministic multiobjective optimization is performed. Since the feasible operating space is highly non-convex, heuristics based techniques that do not require gradient information are used to generate the Pareto (trade-off) surface. Heat integration for the WGS reaction section of the process is necessary for generating steam required as reactant. ASPEN energy analyzer was employed to generate different heat exchange networks that were combined in a supersturcture. Then MINLP compliant process synthesis capability was employed to obtain the optimal design of the this particular heat exchange network synthesis problem.

See more of this Session: Modeling for Process Optimization
See more of this Group/Topical: Process Development Division

Integrating Safety Issues in Conceptual Design of Solvent Processes

Suhani Patel, Artie McFerrin Department of Chemical Engineering, Mary Kay O'Connor Process Safety Center, College Station, TX

Dedy Ng, Artie McFerrin Department of Chemical Engineering, Mary Kay O'Connor Process Safety Center, College Station, TX

M. Sam Mannan, Artie McFerrin Department of Chemical Engineering, Mary Kay O'Connor Process Safety Center, College Station, TX

Integration of safety aspects in process design has become of utmost significance in light of many chemical process incidents that occur frequently. Solvents are commonly applied in various processes and operations but at the same time they are extremely hazardous materials. Many incidents involving fire and explosion events have been escalated because of large amounts of flammable solvents being used in the process. Toxicity hazards are also of concern to many workers in the plant and nearby local residents. Safety can be integrated into the design of solvent processes by means of an optimization formulation. While maximizing profits and satisfying constraints on product quality requirements, additional constraints can be placed on several safety requirements such as a maximum permissible release amount, maximum concentration levels reached upon release of material and risk associated with fire/explosion/toxic events. In this work, a simple case study on liquid -liquid extraction has been shown and the effects of applying such safety constraints on the different process variables have been assessed. Issues relating to the probable reduction in productivity at the cost of safety will be described as well.

See more of this Session: Modeling for Process Optimization
See more of this Group/Topical: Process Development Division

Evaluation of Optimal Control Profiles for Electro-Organic Synthesis Processes Using Batch Electrochemical Reactor

Vijayasekaran Boovaragavan, Chemical Engineering, Tennessee Technological University, Cookeville, TN

Ravi N. Methekar, Energy, Environmental and Chemical Engineering, Washington University in St. Louis, St. Louis, MO

Mounika Arabandi, Chemical Engineering, Tennessee Technological University, Cookeville, TN

Venkat Subramanian, Energy, Environmental and Chemical Engineering, Washington University in St. Louis, St. Louis, MO

Ahmed Basha Chiya, Central Electrochemical Research Institute, Karaikudi, India

Previous contributions [1-3] in batch operations of electro-organic synthesis involve the assumption that the electrochemical reactions are taking place with very fast surface electrochemical transformation. Likewise, the separation of total current into faradic and non-faradic current densities is also greatly ignored. As it is the important factor for dynamic models to account for surface concentration dynamics [4], this talk provides a new profiles for the optimal time-varying potential control by accounting the this phenomena. The main feature of the proposed strategy is that the tasks to account surface dynamics and non-faradaic current are fulfilled all together. Besides, in the present work the results of applying the optimal control theory with those resulting from the best steady controls are also compared to realize the potential benefits of optimal control policy and its evaluation. Some industrially important electro-organic processes are considered for the demonstration [5]. Optimal time-varying electrode potential profiles are obtained using the control vector parameterization (CVP).

The results show improvements in the final yield of the products up to tens-fifteen percent in comparison with constant-steady mode of operation. Hence, it shows economical interest of dynamic operating mode. These results are hopeful and put forward the use of this methodology to other reactor-reaction systems. Once the optimal operation policies are obtained, these can be implemented by designing appropriate controllers.

References

1. R. Bakshi and P.S. Fedkiw, *J. Appl. Electrochem.*, **23**, 715 (1993).

2. F. Fournier, M. A. Latifi and G. Valentin, *Chem. Eng. Sci.*, **54**, 2707 (1999).

3. X. G. Zhou, X.S. Zhang, X. Wang, Y.C. Dai and W.K. Yuan, *Chem. Eng. Sci.*, **56**, 1485 (2001).

4. B. Vijayasekaran and C. Ahmed Basha, *Electrochim. Acta.*, **51**, 200 (2001).

5. R.D. Little and N.L. Weinberg, *Electroorganic synthesis*, Marcel Dekkar, New York, 1991.

See more of this Session: Modeling for Process Optimization
See more of this Group/Topical: Process Development Division

Optimal Scheduling for Ethylene Cracking Furnace System

Chaowei Liu, Dan F. Smith Department of Chemical Engineering, Lamar University, Beaumont, TX

Qiang Xu, Dan F. Smith Department of Chemical Engineering, Lamar University, Beaumont, TX

Chuanyu Zhao, Dan F. Smith Department of Chemical Engineering, Lamar University, Beaumont, TX

Industrial ethylene cracking process employs multiple cracking furnaces in parallel to convert various hydrocarbon feedstocks to smaller hydrocarbon molecules, where mostly are ethylene and propylene. Ethylene cracking operation is also a typical semi-continuous dynamic process, where coking phenomena always accompanies with cracking reactions. Because of coking, a cracking furnace has to be periodically shut down for maintance to restore its performance. Given data of multiple feed characteristics, different furnace performances, various product prices and manufacturing costs, the trade-off between the productivity and cleanup cost gives rise to the optimization opportunities to the entire cracking system for achieving the best economic performance through operation scheduling. To improve the ethylene furnace performance, it is also necessary to consider multiple products, such as propylene and ethylene productions simultaneously.

In this paper, a multi-product considered MINLP model has been developed to address a cyclic scheduling strategy for the best performance of ethylene cracking furnace system. Compared to previous studies, the new developed model has more capabilities to address operation profitability of multiple feeds cracked in multiple furnaces with multiple products. The identified schedule considers the total productivity, maintenance cost, production loss due to the unit shutdown and re-startup, and production yields of major products. In this MINLP model, the batch processing time of a feed processed in the same unit may be different during one cycle operation to maximize the profitability. Meanwhile, the scheduling model can inherently avoid simultaneous cleanups for multiple furnaces, and could maintain the downstream ethylene and propylene production ratio within a specified range, thus results in a more applicable and more stable production schedule for the plant. In this paper, the inherent disturbance due to furnace shutdown will be evaluated. The conceptual design and additional scheduling strategy to completely iron out such disturbances are also proposed. With production yield models from real plant data, a case study has demonstrated the efficacy of the developed methodology.

See more of this Session: Modeling for Process Optimization
See more of this Group/Topical: Process Development Division

How Much API Is Too Small for High Throughput Screening

Shella L. Tameze, Analytical Research and Development, Solid State Characterization Group, Bristol-Myers Squibb, New Brunswick, NJ

Victor Rosso, Pr&d, Bristol-Myers Squibb, New Brunswick, NJ

Anisha Patel, Ar&d, Bristol-Myers Squibb, New Brunswick, NJ

Beth Sarsfield, Ar&d, Bristol-Myers Squibb, New Brunswick, NJ

Shawn Yin, Ar&d, Bristol-Myers Squibb, New Brunswick, NJ

Boris Shekunov, Ar&d, Bristol-Myers Squibb, New Brunswick, NJ

Crystallization from solution is by far the most frequently used method to generate solid materials. Recently, several innovative crystallization tools have been reported, including high throughput screening (HTS) which has been fairly well integrated into pharmaceutical development. The emergence of this technology was largely fueled by the increasing drive to speed pharmaceutical development and yet capture solid form diversity as early and as thoroughly as possible. HTS is especially desirable, not only because it can be used to perform a few thousand crystallizations per week (almost completely automated), but also because it can achieve these results with only a few milligrams of API. HTS is a vital tool in process development as it is used to screen for conditions enabling purification of intermediates and API by crystallization and API form control.

HTS is generally conducted in a 96-well plate system in which a solution of the desired solid is dispensed using automated liquid handling systems. The amount of API reported per well for most HTS lies in the range of 0.25-10 mg. In this presentation, HTS was successfully conducted achieving crystallization with as little as 0.1 mg of API in a well. We will also report that even at a sample loading as low as 0.05 mg of API, a high quality PXRD pattern can be obtained. This illustrates the utility of HTS especially in the early stages of process development when API availability is limited.

See more of this Session: Modeling for Process Optimization
See more of this Group/Topical: Process Development Division

AUTHOR INDEX

AUTHOR INDEX